D0559395

# Ecology and environmental planning

# Ecology and environmental planning

John M. Edington

*Director, Environmental Studies, and Senior Lecturer*
*Department of Zoology,*
*University College, Cardiff, Wales*

and

M. Ann Edington

*Lecturer, Department of Zoology,*
*University College, Cardiff, Wales*

London
Chapman and Hall

A Halsted Press Book
John Wiley & Sons, New York

First published 1977
by Chapman and Hall Ltd
11 New Fetter Lane, London, EC4P 4EE

© J. M. Edington and M. A. Edington 1977

Printed in Great Britain by
J. W. Arrowsmith Ltd., Bristol,
BS3 2NT, England

ISBN 0 412 13300 8

All rights reserved. No part of this book
may be reprinted, or reproduced or utilized in
any form or by any electronic, mechanical or
other means, now known or hereafter invented,
including photocopying and recording, or in
any information storage and retrieval system,
without permission in writing from the
Publisher

Distributed in the U.S.A.
by Halsted Press, a Division
of John Wiley & Sons Inc., New York

**Library of Congress Cataloging in Publication Data**

Edington, J M
Ecology and environmental planning.

Bibliography: p.
Includes index.
1. Environment policy. 2. Ecology. 3. Land use-Planning. I. Edington, M. A., joint author. II. Title.
HC79.E5E323 1977 301.31 77-23076
ISBN 0-470-99262-X

301.31
E23

# Contents

79-1970

# Acknowledgements

We would like to thank Professor D. Bellamy, Department of Zoology, University College, Cardiff, for providing facilities in his department during the preparation of this book. We have drawn freely for examples on a series of research studies in applied ecology, carried out at University College, Cardiff. This programme was made possible by financial support from Glamorgan County Council, the Welsh Office, the Natural Environment Research Council and the Joint Science/Social Science Research Council. Professor J. C. Ene, Department of Biological Sciences, University of Benin, Nigeria provided facilities for the study reported in Chapter 10. We are also grateful for technical assistance from Mrs. Cynthia Carrell and Mrs. Glenys Mackintosh who prepared the final typescript, and Mrs. Gaye Evans and Mrs. Pam Davies who assisted with the preparation of some of the maps and diagrams for the regional case histories.

# 1 Ecology and environmental planning

In its broadest sense environmental planning is an attempt to balance and harmonise the various enterprises, which man, for his own benefit, has superimposed on natural environments. Although these enterprises are often designed to complement one another, possibilities for conflict and imbalance can arise at many levels. On the global scale, some observers fear that our present commitment to industrial growth will lead to increasing environmental damage and resource depletion, and result eventually in the collapse of society (Meadows, *et al.*, 1972; Goldsmith *et al.*, 1972). Economists see an equal danger in the incomplete industrial growth characteristic of many developing countries. They point out that industrialisation at this level frequently creates problems of rural depopulation and loss of food production, without sufficiently increasing prosperity to allow food supplies to be purchased from other countries (Mountjoy, 1975). Whatever the validity of these somewhat pessimistic assessments, and certainly there are some dissenting voices (Maddox, 1972), they are not the subject of this book. Here it is our intention to concentrate on the prospects of successfully integrating different enterprises at the more restricted, regional level.

Any change of land use is a potential source of disruption and the task of every planner is to supervise these rearrangements so that the disharmony they cause is reduced to a minimum. In this task he must make use of two basic concepts. One is that every enterprise has its own *site requirements* in relation to natural features of the landscape. Thus for example, geological formations with poor load-bearing properties are usually considered unsuitable for heavy industrial installations; active river flood-plains are unsuitable for housing and areas with a high incidence of fog are unsuitable for major airports. The second basic idea is that enterprises brought into proximity to one another are likely to show different *degrees of compatibility*. For example, the siting of a residential area alongside a major airport or a steel works would normally be precluded on the grounds of noise and pollution respectively. In other words these combinations have a low compatibility. By contrast, many water-supply reservoirs also cater for activities such as angling and sailing, and in this case the two enterprises, water supply and recreation are highly compatible.

Of course, advances in technology may eventually remove some of the constraints. We now know that new types of foundation construction may allow heavy buildings to be erected on unstable strata. Quieter jet engines could reduce the noise disturbance zone around airports, and improved methods of treating effluents discharged to the atmosphere could allow housing to be sited closer to industry.

Public opinion helps to dictate what functions can be considered compatible and public opinion is no less changeable than technology. In Britain there are at present many people who feel that coniferous plantations should be excluded from areas of scenic beauty particularly in those areas where the deciduous hardwood trees formed the natural vegetation, but there is no certainty that this attitude will persist when the landscape possibilities of mature plantations become more evident.

Planners are obliged to review the current state of these various relationships whenever they are called on to evaluate a proposal for a change in land use. For this purpose not only do they need to be attuned to public opinion, but they also need evaluations from a wide variety of specialist disciplines which may include geology, ecology, medicine, sociology, economics and engineering. Information from specialists can be sought on an *ad hoc* basis or be arranged within the format of an 'Environmental Impact Analysis', a method widely adopted in America (Johnson, 1975). By marrying assessments about local impacts with expressions of national policy, the planner hopes to arrive at the right decision about rejecting, accepting or modifying any new development. It must be recognised, however, that no country has an entirely comprehensive planning system. There is rarely effective liaison between urban and industrial planning on the one hand and rural planning on the other.

## The ecological contribution

We must consider next the nature of the specifically ecological contribution in planning. Ecology is that part of biology which is concerned with the external relationships of organisms (as opposed, for example, to a study of their internal structure). These relationships include not only reactions to physical and chemical factors in the environment but also interactions with other species. In some circumstances it is not difficult to see how information on these matters can have a direct bearing on planning. If, for example, it is proposed to site a coal-washery on a trout river the knowledge that trout can tolerate a maximum of 80 mg/l suspended solids is directly relevant to the question of whether the fishery and the industrial installation can coexist.

In other circumstances connections are not so immediately obvious; often the important information concerns the links between species. The hazards associated with industries using mercury compounds were only fully appreciated when it was discovered that aquatic micro-organisms could transform relatively innocuous inorganic mercury wastes into the dangerous methyl form. By passing through food chains and accumulating in organisms such as fish eaten by man, methyl mercury has caused serious poisoning in a number of communities which rely heavily on seafoods.

Although any involvement by an ecologist in planning issues is in a sense an involvement with the 'ecology' of man, most ecologists would deny their competence to comment on *Homo sapiens* quite as comprehensively as on any other species. After all, many facets of human existence are already covered by disciplines such as medicine, sociology and economics, developed for specific purposes. Whilst it is essential that ecologists should collaborate with specialists in these disciplines, it is inappropriate to represent these subject areas as aspects of ecology.

Two other sources of confusion need to be mentioned. Ten years ago an 'ecologist' was quite unambiguously a scientist interested in analysing the environmental relationships of living organisms. More recently the term has come to be used in a second sense, not to identify a practitioner in a particular scientific discipline but rather to indicate a philosophical attitude, often an attitude involving a commitment to conservation and an antipathy to development. Whilst this second kind of 'ecologist' certainly has a contribution to make to planning, the planner might do well to classify this input in the 'public opinion' rather than the 'scientific' sector. A related problem is that ecology is frequently confused with conservation. There is a widespread misapprehension amongst planners (and possibly some ecologists) that an ecological evaluation consists solely of identifying and grading sites of biological conservation interest. This is certainly too restricted a view. In Britain it probably derives from the fact that for many years the Nature Conservancy was the only ecological agency to which planners had direct access, and inevitably this contact resulted in emphasis being given to conservation considerations at the expense of wider ecological issues.

For the purpose of participating in environmental planning, ecologists need to organise their information, not simply according to the traditional subject areas within ecology (species ecology, population dynamics, community structure) but also in terms of the enterprises that are the common currency of planning. The present account is arranged accordingly, first to examine the implications of particular types of development (Chapters 2–6) and then through a series of case histories (Chapters 7 – 10) to look at the ecological problems associated with the application of general planning strategies to whole regions.

# 2 Rural land use

Rural land is required to fulfill a wide range of functions and of these, the requirements of crop production and water collection can be considered as the most basic. Increasingly it is being realised that the management of rural land must also cater for biological conservation and public recreation. This approach has arisen in the more developed countries as a reaction to the increasing use of land for urban and industrial purposes, and for intensive agriculture. Although, in theory, rural land uses can be combined in a variety of ways, possible combinations show different degrees of compatability. Thus, whilst a forest managed for timber production can accommodate a wide range of recreations within its boundaries, this is not the case with a lowland farm. Farmland, however, may form an essential background to those sight-seeing activities which involve the aesthetic appreciation of landscape.

A significant issue is the extent to which conservation can be combined with crop production or recreation, compared with the extent to which it needs to be catered for on sites set aside for this particular purpose.

Apart from these questions of compatibility, the wide range of rural land uses can be graded according to their site requirements. Some forms of recreation such as climbing, caving and skiing have very specific site requirements in relation to the natural physical features of the environment. As will be shown this is also an important consideration in the allocation of land to different crops. Although ideally, rural land should be organised comprehensively to take account of all these aspects, it is probably true to say that no country in the world has a planning system which caters adequately for rural land use.

## Crop production

There is a large body of information available on the site requirements of different crops. Such information has a direct relevance to rural planning but is rarely fully utilised for this purpose. Too often the spatial arrangement of crops is allowed to owe more to antiquated land-tenure patterns and interdepartmental rivalries, than to the variables such as soil type, topography and aspect which are likely to influence productivity.

The balance between hill sheep farming and forestry in the uplands of Britain provides a striking example of some of these problems. At altitudes below the natural tree line in Britain (550 – 600 m) most hill areas are capable of supporting either enterprise. However, each operation faces a different set of site limitations on different sections of the hill profile. As far as forestry is concerned the plateaux tops and the steep hill sides can each present problems. The plateaux tops are usually covered with a mixture of deep peats and 'gleyed' soils, the latter waterlogged close to the surface. Trees do not grow well on deep peats, and, although they can establish themselves on waterlogged sites, the root mat is likely to be so shallow that when the trees exceed about 10 m in height they are liable to be blown over by the wind. The steep slopes may also include areas of poor forest growth. Here there is no impedance of drainage and it is the free movement of water down the slope that is the main problem because it causes the leaching of plant nutrients from the soil. These problems leave the hill crests and the lower slopes as the optimum habitats for forest growth. The hill crests are sufficiently well-drained to avoid waterlogging but water movement is not active enough to produce nutrient depletion. The lower gentle slopes are typically covered by deeper, richer, brown earth soils and have the advantage of receiving the nutrients washed from the upper slopes.

Some of the same site limitations affect the sheep farmer, but there are also important differences. The key areas for a hill sheep farm are undoubtedly the lower slopes and valley-bottom land with their good soils and sheltered position. These are the obvious sites for the farmhouse and farm buildings, for improved pastures and fields of fodder crops. Probably next in order of usefulness are the gleyed soils on the plateau top. These have proved surprisingly amenable to pasture improvement, usually by a combination of reseeding, fertiliser application and controlled grazing in fenced enclosures. The farmer faces the same difficulties as the forester in making use of the deep peats. The steep slopes also present problems. They provide relatively poor grazing for sheep and are usually regarded as incapable of significant improvement by ploughing and reseeding, because of the difficulty of using machinery on them. A slope of 13° is usually regarded as the typical working limit for an ordinary farm tractor, and 20° for a caterpillar tractor (Curtis *et al.*, 1965).

It is possible on this basis, to define a forest/farm landscape which incorporates the site priorities and site limitations of each enterprise. Thus the valley-bottom land and plateau top might be allocated to farming, the hill crests to forestry, and the steep-slope land variously distributed to make up units of a manageable size. This last consideration is particularly important with forestry because of the high cost of fencing and providing access roads for a large number of small plantations. In contrast with many parts of the world where there is a seasonal concentration of rainfall, soil erosion is not a major hazard in upland Britain and no land need be excluded from development on this basis.

The application of these criteria to an actual hill area in mid-Wales would produce the pattern shown in Fig. 2.1*a*. When this is compared with the pattern which has actually emerged (Fig. 2.1*b*) it becomes clear that the allocation of land to farming and forestry is not made primarily on an ecological basis. Apart from the limitations imposed by previous ownership patterns, reflected in the angular nature of many of the boundaries, the large block of common land has been a major impediment to land improvement. These and similar difficulties of planning cropland on an ecological basis appear in some form or another in most parts of the world.

## Conservation

### *The prospects of accommodating wildlife conservation in crop-producing areas*

There are many reasons for being pessimistic about the scope for conservation on farmland or in commercially managed forests. The characteristics of such crop-producing areas in terms of physical structure and plant species diversity are often poor substitutes for the habitats they replaced and many animal species are unable to survive the transition.

In Britain, for example, where large areas of natural oak-dominated woodlands have been replaced by moorlands and by coniferous plantations, very few of the typical insects of broadleaved woodland persist in the new habitats. Other groups such as mammals and birds fare a little better, and although few oak-woodland songbirds can survive on bare moorland, about half the species are found in mature conifer plantations (Table 2.1).

Habitat changes are not the only hazards to animals. Many representatives of natural faunas which survive on croplands are actively persecuted as pests or suffer inadvertently from chemical control measures directed at other organisms. These influences are well illustrated by the problems which face the surviving remnants of the original prairie fauna in North America, now living in an environment dominated by stock rearing and grain production. Poisoning and trapping campaigns have been directed against the coyote (*Canis latrans*) because of its attacks on farm animals, and against prairie dogs (*Cynomys* spp.), ground squirrels (*Citellus* spp) and gophers (*Geomyidae*) because of the damage they cause to crops and grazing land (Fig. 2.2). In the case of the prairie dog these measures, coupled with

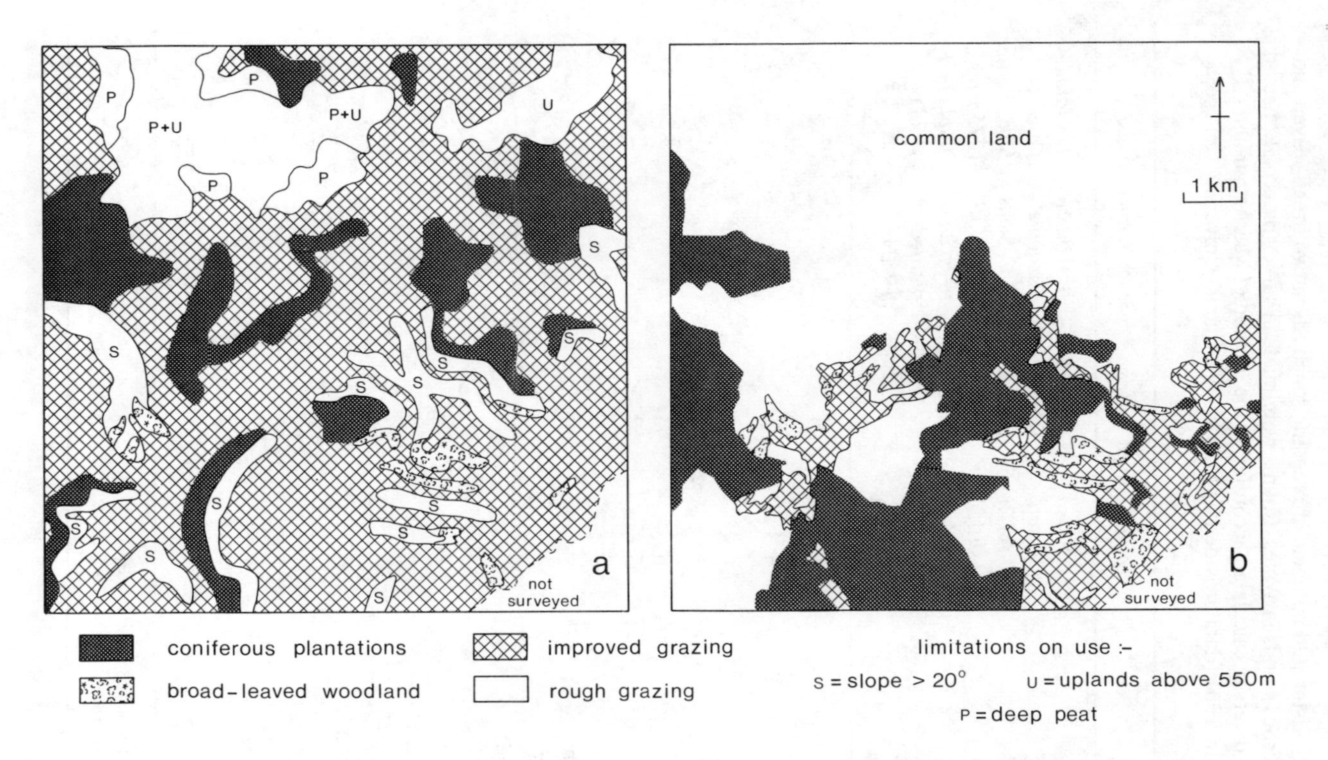


*Fig. 2.1* The allocation of land to hill-farming and forestry in an area in mid-Wales. (*a*) The optimum pattern indicated by ecological considerations. (*b*) The pattern as it is (after Brummage, *et al.* 1977)

*Table 2.1* Estimated numbers of breeding songbirds in two broad-leaved and two coniferous woods in Wales. Counts were made in survey plots of 12 hectares, indices of diversity calculated according to the method of MacArthur and MacArthur, 1961 (data provided by Miss M. W. Adams, University College, Cardiff).

| | *Broad-leaved woods* | | *Coniferous woods* | |
|---|---|---|---|---|
| | *1. Coed y Croftau Oak/Birch/ Ash (275 trees/ha relict)* | *2. Coed y Rhygen Oak/Birch (375 trees/ ha relict)* | *3. Tintern Corsican Pine/Scots Pine (450 trees/ha planted 1923)* | *4. Hafod Fawr Sitka Spruce (275 trees/ ha planted 1928)* |
| Chaffinch | 7 | 4 | 4 | 7 |
| Wren | 13 | 15 | 34 | 9 |
| Robin | 7 | 4 | 3 | 4 |
| Dunnock | – | – | 2 | – |
| Goldcrest | 2 | 4 | 9 | 31 |
| Coal tit | 6 | 3 | 7 | 4 |
| Blue tit | 5 | 2 | – | – |
| Great tit | 6 | 2 | – | – |
| Marsh tit | – | – | – | – |
| Nuthatch | 3 | – | – | – |
| Treecreeper | 1 | 1 | – | – |
| Blackbird | 3 | – | – | – |
| Song thrush | – | – | – | – |
| Mistle thrush | – | – | – | 3 |
| Wood warbler | 3 | 3 | – | – |
| Willow warbler | 4 | 10 | – | – |
| Blackcap | – | – | – | – |
| Chiffchaff | – | – | – | – |
| Pied flycatcher | 6 | 7 | – | – |
| Redstart | 2 | 2 | – | – |
| TOTAL | 68 | 57 | 59 | 58 |
| Number of species | 14 | 12 | 6 | 6 |
| Index of diversity | 2·47 | 2·21 | 1·31 | 1·37 |

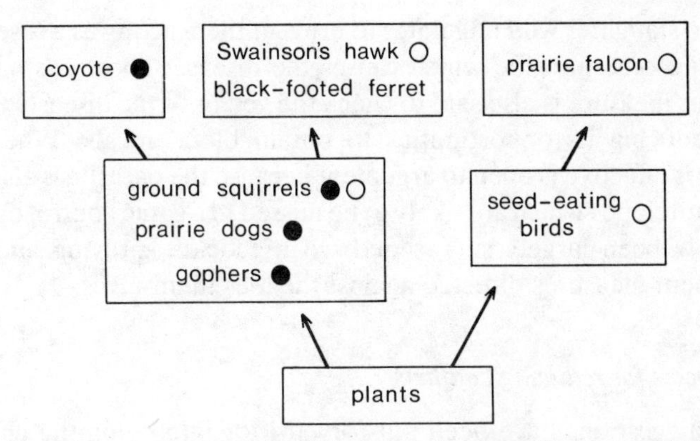


*Fig. 2.2* Some features of prairie food chains showing: (i) species which have been controlled as pests ●; (ii) species which have been affected indirectly by pesticides ○. Two other typical prairie species, the bison (*Bison bison*) and its predator the plains wolf (*Canis lupis*) were virtually eliminated during the initial exploitation of the region

habitat changes, have reduced its range to a few protected localities. This decline has brought an important predator of the prairie dog, the black-footed ferret (*Mustela nigripes*), to the verge of extinction (Clark, 1976). At least two predatory birds, the prairie falcon (*Falco mexicanus*) and Swainson's hawk (*Buteo swainsoni*) are known to have declined as a result of the routine use of mercury-based fungicides and organo-chlorine insecticides as seed dressings. These materials accumulated in seed-eating rodents and birds and were passed on to the predators via the food chain (Fyfe *et al.*, 1969; Fimriette *et al.*, 1970).

The control of insect pests in North American forests has also had some deleterious effects on natural communities. The coniferous forests along the eastern seaboard have suffered serious damage as a result of attacks by the spruce budworm (*Choristoneura fumiferana*) and in an attempt to control these outbreaks, millions of hectares of forest have been sprayed with insecticides. At first DDT was used but this proved to have very damaging effects on stream and river faunas, both fish and invertebrates. Subsequently a change was made to the organo-phosphorus insecticide phosphamidon. This proved virtually harmless to fisheries but in the concentrations initially used caused serious mortality amongst song birds (Rudd, 1964; Dagg, 1974).

Another group of animals which has suffered as a result of agriculture are the species implicated in disease cycles involving farm stock. Recently it has been thought necessary to eliminate badgers (*Meles meles*) from a number of farms in the west of England to reduce the threat of them transmitting tuberculosis to cattle. In many parts of southern Africa it has been a common practice to slaughter wild ungulates to prevent them acting as a reservoir for the trypanosome parasite which causes the disease, known as nagana, in cattle. This measure is also said to check the vector of the disease, the tsetse fly, by reducing its opportunities to obtain blood meals. Whether this approach is effective is open to argument because the parasite is also carried by small antelopes which are likely to be missed in a game control operation. It has now been largely superseded by insecticide spraying and habitat management measures directed against the flies themselves.

### *The prospects for reducing conflicts*

Various suggestions have been put forward for improving the chances of wildlife surviving on crop land. Obviously one possibility is to leave more fragments of natural habitats to serve as refuges within farms or forests. In Britain there have been a number of exercises carried out on different farms in an attempt to assess the concessions that might reasonably be made in the intests of wildlife (Barber, 1970; Dorset Naturalists' Trust, 1970). On lowland farms, the main difference between the agriculturalist and the conservationist lies in his attitude to hedgerows. The conservationist sees the hedgerow as a valuable habitat for wildlife, particularly birds, whereas the farmer often sees it as an annual commitment in terms of maintenance and as an impediment to the operation of his farm machinery. In these circumstances it is difficult to imagine farmers retaining hedgerows which involve them in a loss of production.

By the same token it would be unreasonable to require farmers to desist from controlling mammal and bird species which are demonstrably causing damage to crops. There are however always areas of uncertainty and close examination of these might lead to decisions in favour of the wild species. On the prairies for example, are the coyote's attacks on farm stock outweighed by its role in the rodent control; are the ground squirrel's attacks on grain outweighed by its predation on harmful insects?

In the field of chemical control there is obviously the possibility of modifying the choice of pesticides and regulating their mode of application to minimise effects on non-target species. There is also the prospect of integrating chemical and biological control methods. One approach to spruce budworm control involves spraying insect-infested trees with both an insecticide and an insect virus. The pathogen so enhances the effect of the insecticide that very low insecticide concentrations can be used, thus greatly reducing the chance of side-effects.

In theory, the ideal way of controlling pests and maintaining natural species in cropping areas would be to use some of these species as control agents. Moth caterpillars and saw-fly larvae are an important part of the diet of wood ants (*Formica* spp.) and experiments in Italy, Germany and the Netherlands have shown that forest trees can be substantially protected from these pests by artificially increasing the number of ant colonies (Franz, 1961). In forest plantations the shortage of natural nesting holes (normally associated with dead trees) sets a limit on insectivorous bird populations. In Europe the erection of nest boxes, which can appreciably increase bird populations, has been widely advocated as a device for controlling insect pests. However the results are ambiguous and there is no certainty that bird populations augmented in this way are capable of checking serious pest outbreaks (Buckner, 1967; Murton, 1971).

### *Game ranching*

In relatively open habitats, cropping systems based on wild ungulates rather than domestic stock are often proposed as a way of simultaneously obtaining an economic yield and conserving wild species.

The exploitation of the saiga antelope (*Saiga tatarica*) in the Soviet Union provides the most striking success story of this kind (Bannikov, 1961). The animal had been over exploited because of the imagined medicinal properties of its horns, and by the beginning of the present century was nearly extinct (Fig. 2.3). However uncontrolled hunting was banned in the 1920's, the herds built up again and a system of controlled cropping was instituted. Each year, the numbers to be taken are worked out on the basis of a twice-yearly population census. The annual harvest can produce as much as 22 000 $m^2$ of leather and 6000 tons of meat. Under this regime total numbers have risen from about 1000 to more than 2 million, and in the process the herds have reoccupied most of their original range (Fig. 2.3).

Without doubt this arrangement successfully combines cropping with conservation. It is however somewhat misleading as a model for other regions because the saiga antelope, with its specially swollen nose for dust filtration and other physiological adaptations, is one of the few ungulate species which could survive on the dry arid steppes of Kazakhstan. There is no real possibility of rearing conventional domestic animals in such conditions. In less extreme situations it becomes appropriate to compare the relative merits of game ranching and conventional stock rearing.

A number of the African savanna habitats are apparently capable of supporting a weight of ungulates which is between three and fourteen times the figure for domestic stock on the same land (Table 2.2). The greater efficiency of the wild species is related to the way in which they have divided up the food resources in the habitat. Different species have become adapted to feed on trees and shrubs at different heights (Fig. 2.4), and even the exploitation of a single grass species is often carried out in a mutually

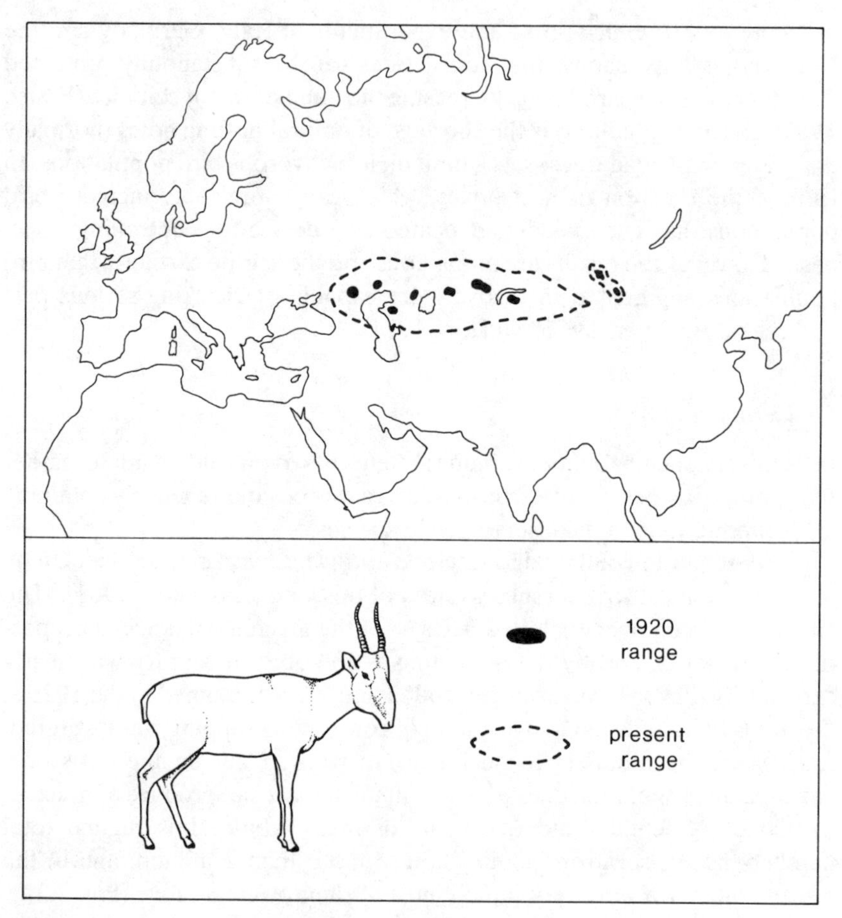


*Fig. 2.3* Extension in range of the Saiga Antelope (*Saiga tatarica*) after the introduction of controlled cropping (after Bannikov, 1961).

*Table 2.2* A comparison of the weights of domestic animals and wild ungulates supported by two East African Savanna habitats (from Talbot *et al.*, 1965)

| *Type of habitat* | *Animals* | *Weight of animals supported kg/hectare* |
|---|---|---|
| Acacia savanna | (a) Cattle, sheep, goats | 19·6 – 28·0 |
| | (b) Wild ungulates | 65·5 – 157·6 |
| Acacia-commiphora bushland | (a) Sheep, goats | 3·7 – 13·5 |
| | (b) Wild ungulates | 52·5 |

*Fig. 2.4* Differences in browsing height in animals of the African wooded savanna: (*a*) giraffe (*Giraffa camelopardalis*), (*b*) elephant (*Loxodonta africana*), (*c*) eland (*Taurotragus oryx*, (*d*) impala (*Aepyceros melampus*), (*e*) steinbok (*Raphicerus campestris*). (After Dasmann 1964)

complementary fashion. It has been shown for example on the Serengeti plains, that the red oat grass (*Themeda triandra*) is cropped by the zebra (*Equus burchelli*), wildebeest (*Connochaetes taurinus*) and Thomson's gazelle (*Gazella thomsonii*), each species modifying the plant cover in a way which assists the feeding of the one following it (Gwynne and Bell, 1968; Bell, 1971).

These observations, coupled with the fact that wild ungulates are unaffected by the parasite which causes nagana in cattle, have led many biologists to argue that game ranching is the best way of using these marginal lands (Huxley, 1961; Dasmann, 1964).

One of the first game ranching schemes was organised on the Doddieburn ranch in the Rhodesian low veldt back in the early 1960's. Essentially the procedure was to calculate the annual production of each of the principal species and to crop the appropriate number of animals (Table 2.3). The cropping was mainly carried out from hides and at night using spotlights, and was claimed to cause little general disturbance to the populations. Since these initial experiments, other game ranching enterprises have been established and by 1967 over 100 cropping licenses had been granted (Pollock, 1969).

Although game ranching may result in wild ungulates being retained on land which would otherwise have been cleared for domestic stock, it departs considerably from the conservationist's ideal of maintaining intact, natural

*Table 2.3* Game populations and yields from a 130 km$^2$ area on the Doddieburn Ranch, Rhodesia (from Dasmann, 1964)

| *Species* | *Estimated numbers* | *Recom-mended crop* | *Weight dressed carcass (kg)* | *Total meat yield (kg)* |
|---|---|---|---|---|
| Impala | 2 100 | 525 | 29 | 15 225 |
| Zebra | 730 | 146 | 116 | 16 936 |
| Steenbuck | 200 | 40 | 5 | 200 |
| Warthog | 170 | 85 | 32 | 2 720 |
| Kudu | 160 | 48 | 102 | 4 896 |
| Wildebeest | 160 | 32 | 118 | 3 776 |
| Giraffe | 90 | 15 | 454 | 6 810 |
| Duiker | 80 | 28 | 9 | 252 |
| Waterbuck | 35 | 7 | 91 | 637 |
| Buffalo | 30 | 5 | 259 | 1 295 |
| Eland | 10 | 2 | 272 | 544 |
| Klipspringer | 10 | 3 | 6 | 18 |
| Bush pig | 10 | 5 | 32 | 160 |

communities complete with all their constituent species, including predators. In all cropping enterprises there is an understandable desire to remove predators, to concentrate on the best meat producers and ultimately to select strains for such characteristics as docility and carcass quality. This is of course the beginning of the process which leads to domestication. In attempting to cater both for meat production and conservation in the same operation it is difficult to escape the conclusion that both functions are being compromised.

### *The design of a world-wide system of reserves*

In view of the difficulties of catering for conservation in crop-producing areas (or for that matter in urban and industrial regions) most biologists have concluded that the needs of conservation can best be met in a system of protected reserves managed primarily for this purpose. Clearly such a system must be designed to reflect the world-wide variation of habitats and communities. Various systems of classification have been suggested. Darwin's contemporary, Alfred Russel Wallace, drew attention to the fact that similar habitats in different continents often support quite different assemblages of animals. For example, the grazing role fulfilled by antelopes in African grasslands is taken over by guinea pig-like rodents in South America and by kangaroos in Australia (Wallace, 1876). Such observations led him to divide the land masses into faunal regions, still generally referred to as Wallace's realms (Fig. 2.5).

The other kind of classification, based on botanical observations, divides the land surface of the world into natural vegetation zones. The maps produced on this basis show the situation not as it is, but as it would have been in the absence of human intervention (Fig. 2.5). The natural patterns of vegetation are determined largely by climatic variables such as rainfall and temperature. Thus the drier hearts of continents are often characterised by grassland because there is an insufficient supply of moisture to allow the development of trees. Similarly in northern continental areas the decrease in temperature and the shorter growing season towards the poles produces a succession of vegetation zones ranging from broad-leaved forest in the south, through coniferous forest, to treeless tundra in the north. Each vegetation type tends to support its own characteristic assemblage of animals, although there are some species which can extend over a number of vegetation zones. In the northern himisphere, bears and wolves are notable examples of this.

In contrast to the terrestrial situations, much less is known about the range of variation in aquatic habitats. Varvious geographical classifications have been proposed for areas of the sea (Ekman, 1953) but the systematic study of inland waters, particularly running waters, is still in its infancy (Luther &Rzóska, 1971).

In any classification of habitats, islands have to be treated separately. Many of them support groups of species which have evolved in isolation and have no counterparts elsewhere. For example, Taiwan is important for its pheasants, Madagascar for its lemurs and the Seychelles group for its unique birds and double-coconut palm or coco-de-mer (Penny, 1968; Swabey, 1970). Island communities have been particularly significant in demonstrating the process of evolution. Indeed it was Darwin's study of the remarkable fauna of the Galapagos Islands, including the giant tortoises and unusual finches, which helped him to formulate the theories advanced in the 'Origin of Species'.

Within continents too, there are habitats such as isolated mountain ranges, deserts and lakes which have many of the properties of islands. In Africa for example the Simien fox (*Simenia simensis*), the walia ibex (*Capre walie*) and the gelada baboon (*Theropithecus gelada*) are all species which are restricted to the Ethiopian Highlands and absent from the mountain ranges in central and southern Africa (Blower, 1968). Similarly Lake Victoria and the two rift valley lakes Malawi and Tanganyika have, by virtue of isolation, their own distinctive assemblages of fish, mollusc and crustacean species (Fryer & Iles, 1972; Beadle, 1974).

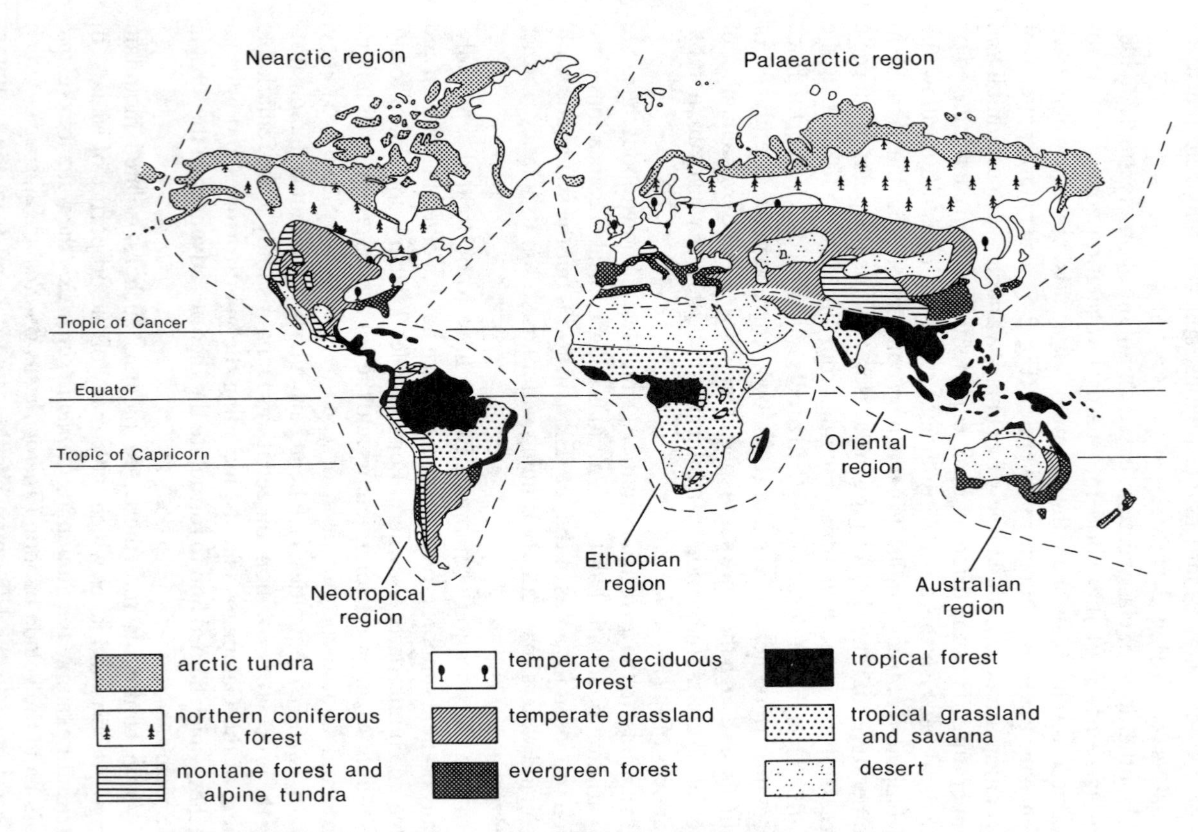


*Fig. 2.5* Map showing Wallace's realms and the world's natural vegetation zones

If these natural distribution patterns of plants and animals were substantially intact it would be a simple matter to devise a world-wide system of representative reserves. In fact, man has disrupted these patterns in many ways, but perhaps most notably by the domestication of plants and animals, and by the dissemination of these favoured species throughout the world. The task of establishing reserves has proceeded nonetheless and throughout the world there are now over 1000 reserve sites which are representative of major community types. Table 2.4 illustrates how the classification of communities has been translated into a reserve system for one particular continent.

As it is rarely possible to find entirely intact and complete communities to establish as reserves, many sites need to be both rehabilitated and continuously managed.

*Rehabilitating natural habitats in reserves and parks*

*Residual problems associated with domestic animals and crop cultivation.* The history of the Serengeti National Park in Tanzania affords a good illustration of the difficulties of disentangling conservation projects from man's pastoral and agricultural activities. When the Park was set up in 1951 it contained substantial herds of cattle owned by Masai tribesmen. Something like 20 000 cattle were being kept on the Serengeti Plains and 100 000 on the Crater Highlands (Fig. 2.6*a*). It was feared that these animals would compete for grazing with the large herds of wild ungulates which the Park was designed to protect (Pearsall, 1957). Accordingly it was decided in 1958 to remove cattle from parts of the Park and to redefine its boundaries. This produced two units, a cattle-free western area designed for conservation and tourism and an eastern unit designated as the Ngorongoro Conservation Area in which the function of cattle herding was to be combined with conservation and tourism (Fig. 2.6*b*). The arrangement has proved fairly satisfactory despite initial concern that the ungulate herds would be inadequately protected during their seasonal migrations into the eastern Conservation Area. (Grzimek & Grzimek, 1964; Fosbrooke, 1972). The exclusion of cattle from the drier, fragile western grazing lands represents a major advance. However, even now the Park falls short of providing the ecological unit necessary to protect migrating herds all the year round. The western boundaries had to be drawn to exclude human settlements and cultivated land, and as a result when the herds move outside the Park at the beginning of the dry season, they cause damage to farmland and render themselves vulnerable to attack by poachers (Fig. 2.7*b*). Thus although the Park is already very large (about 13 800 $km^2$), it would need to be extended even further to establish a self-contained ecological unit.

*Table 2.4* The representation in reserves and parks of the main biological community types in Africa south of the Sahara

| *Community type* | *No. of reserves* | *Example* | *Area ($km^2$)* |
|---|---|---|---|
| *Tropical Rain Forest* | | | |
| 1. Congo rain forest | × | Salonga National Park, Zaire | 36 000 |
| 2. Guinean rain forest | × | Tai Faunal Reserve, Ivory Coast | 4 250 |
| *Tropical Savanna* | | | |
| 3. West African wooded savanna | ×× | Yankari Game Reserve, Nigeria | 2 100 |
| 4. East African wooded savanna | ×× | Luangwa National Park, Zambia | 15 500 |
| 5. Congo wooded savanna | × | Upemba National Park, Zaire | 1 000 000 |
| 6. South African wooded savanna | ×× | Mkuzi Game Reserve, South Africa | 250 |
| 7. West African thorn-tree/grass savanna | × | Boucle du Baoulé National Park, Mali | 5 430 |
| 8. East African thorn-tree/grass savanna | ×× | Serengeti National Park, Tanzania | 13 800 |
| 9. Karroo thorn-tree/grass savanna | × | Mount Zebra National Park, South Africa | 46 000 |
| 10. Cape sclerophyll | × | Addo Elephant Park, South Africa | 200 |
| *Desert* | | | |
| 11. Namib | × | Namib Desert Park, Namibia | 3 000 |
| 12. Kalahari | × | Kalahari Gemsbok National Park, South Africa | 9 600 |
| *Mountain and Highland* | | | |
| 13. Ethiopian highlands | × | Simien Mountains National Park, Ethiopia | 1 000 |
| 14. Guinean highlands | × | Kimbe River Game Reserve, Cameroon | 40 |
| 15. Central African highlands | ×× | Virunga National Park, Zaire | 8 100 |
| 16. South African highlands | ×× | Golden Gate Highlands National Park, South Africa | 42 |

× = few, ×× = many

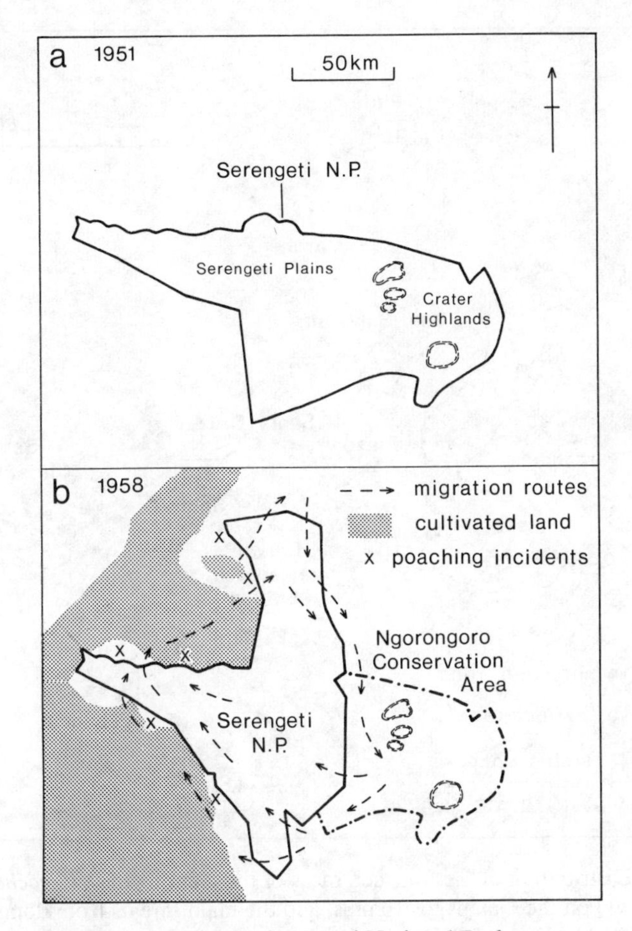


*Fig. 2.6* Boundary changes in the Serengeti National Park

Problems arising from crop cultivation and domesticated animals arise in another form in the Galapagos National Park, which was designated in 1959 to protect the islands' unique plant and animal populations.

Some of the unusual plant assemblages such as the groves of the woody composite *Scalesia* and the shrub *Miconia* have been partially replaced by crop plants such as castor-bean (*Ricinus communis*) and guava (*Psidium guajava*) (Schofield, 1973). Even more significant has been the impact of domestic stock. Goats and pigs introduced by settlers now range freely over many of the islands (Fig. 2.7), and interfere with the natural flora and fauna in a variety of ways. (MacFarland *et al.*, 1974a,b). By their intensive browsing activities, the goats have so reduced the vegetation cover that it has become more difficult for the land iguanas (*Conolophus* spp.) to escape from bird predators. The goats also reduce the supply of plant food available to

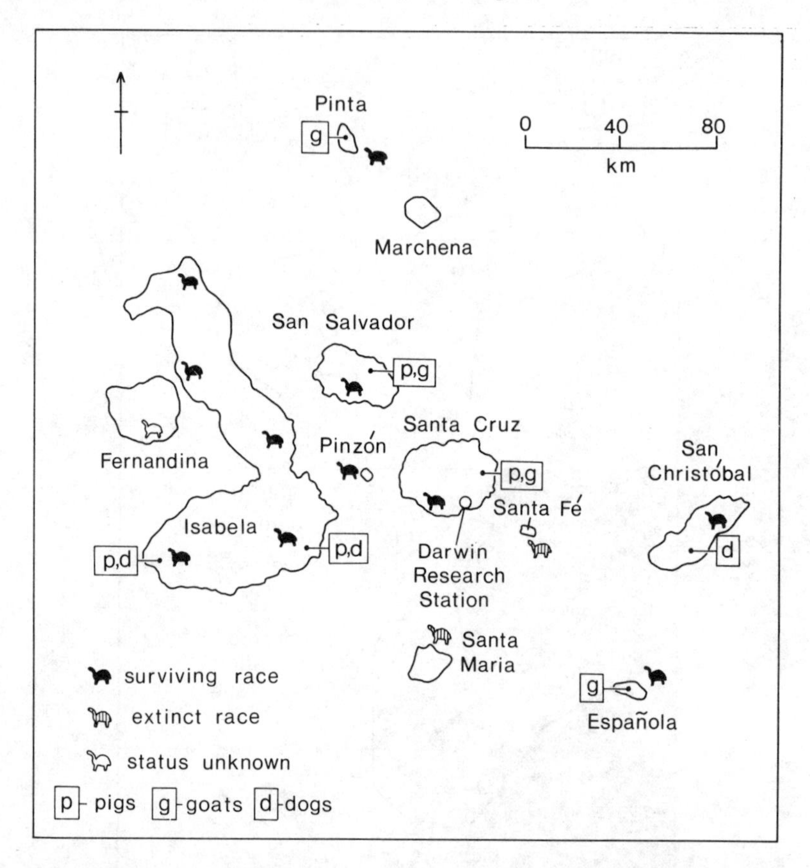


*Fig. 2.7* Distribution of the fourteen races of the giant tortoise (*Geochelone elephantopus*) on the Galapagos Islands, and the main threats from domestic species (after MacFarland & Reeder 1975)

the giant tortoises (*Geochelone elephantopus*). Free ranging pigs on some of the islands pose another threat to the tortoises by digging up eggs and eating the young.

The iguanas and tortoises are regarded as particularly important in the way they illustrate the principles of evolutionary divergence in isolated populations. The small island of Santa Fé has its own species of land iguana which is separate from the species on the other islands. Even more strikingly the giant tortoise has diverged into 14 different races (two or possibly three of which are now extinct) all occupying different islands or different sections of the largest island of Isabela (Fig. 2.7).

Away from areas of human settlement the obvious protective measure is to try and eliminate the goats and pigs. The clearance of goats from two small islands Santa Fé and South Plaza (a very small island off Santa Cruz) has greatly benefited the land iguanas. However on the larger islands the task is more difficult. During 1971 and 1972 the removal of 18 000 goats from Pinta still only reduced the population by a half. With the pigs some control of numbers has been achieved on San Salvador and Santa Cruz but there is little prospect of making substantial inroads into the enormous populations roaming on the large main island of Isabela. Fortunately there is an alternative approach to the pig problem. It has been found that walls constructed from lava stones keep the pigs out of tortoise breeding areas and prevent them from eating the tortoise eggs and young (Table 2.5). Unfortunately the walls give no protection against the dogs which sometimes get into the enclosures and cause heavy mortality. The only certain way to avoid these early hazards is to hatch the tortoise eggs in incubators and rear the young in captivity until they are large enough to resist attacks from predators. This rearing and release programme is organised by the Darwin Research Station on Santa Cruz (MacFarland & Reeder, 1975).

*Table 2.5* Effect of protecting tortoise nests from pig-predation (from MacFarland *et al.*, 1974b)

| *Location* | *Race* | *Nesting season* | *No. of nests protected* | *No. of protected nests destroyed* |
|---|---|---|---|---|
| Santa Cruz | proteri | 1970–71 | 46 | 1 |
| | | 1971–72 | 115 | 0 |
| San Salvador | darwini | 1970–71 | 38 | 0 |
| | | 1971–72 | 31 | 0 |
| Isabela | vicina | 1971–72 | 32 | 0 |

The examples of the Serengeti and Galapagos Parks serve to illustrate that disruptive human influences are not automatically halted by the establishment of a park or reserve. In many island reserves domestic animals and other introduced species create a continuing management problem. In continental areas difficulties are more likely to result from some compromise over boundaries which prevents the reserve from functioning as a natural unit.

*Problems of overpopulation in reserves.* In some reserves problems arise because one of the principal species increases in numbers to such an extent that it threatens to cause serious habitat damage. This phenomenon has been observed in protected populations of deer, moose, seals, hippos and elephants. The interpretation of these situations obviously has important implications for the design and management of reserves.

Overpopulation in deer is relatively easy to interpret. It is usually associated with situations where natural predators such as wolves have been eliminated, supposedly in the interests of domestic stock or game animals. The story of the mule deer (*Odocoileus hemionus*) on the Kaibab plateau in Arizona provides an interesting example. This area was made into a game refuge in 1906 and, as was then the usual practice, a predator control programme was instituted which exterminated wolves and drastically reduced the numbers of other predators such as coyotes, mountain lions and bobcats. The deer populations now immune from the controlling effect of predation, increased dramatically from 4000 to a peak of 100 000 in 1924, and, in the process, so damaged the vegetation that the herds could no longer find sufficient food. Starvation set in and caused a dramatic collapse, with something like 90 000 deaths over a period of two years (Fig. 2.8).

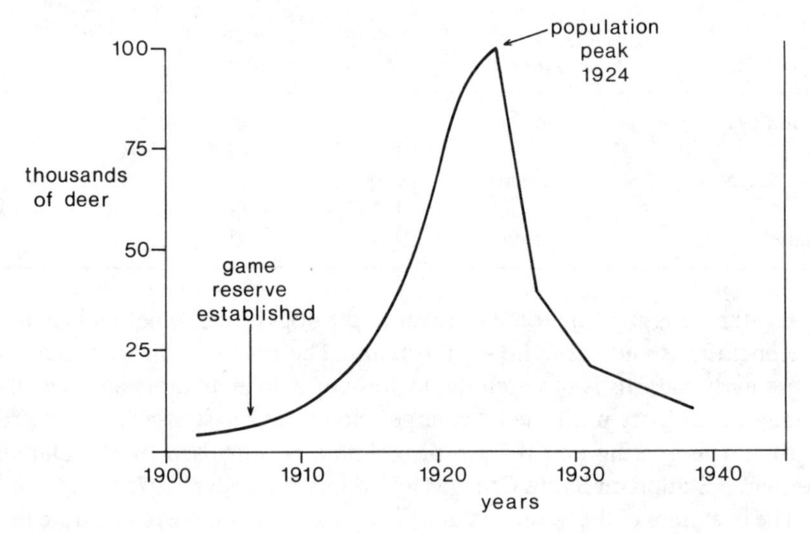


*Fig. 2.8* The population irruption of mule deer (*Odocoileus hemionus*) on the Kaibab plateau, Arizona (after Rasmussen 1941)

A similar demonstration of the controlling action of wolves has been provided by their interaction with moose (*Alces alces*) on Isle Royale, an island in Lake Superior. Moose colonised the island at the beginning of the century and in the absence of major predators the population grew rapidly, seriously damaged its own food supply (mainly the foliage of shrubs) and within 30 years had crashed dramatically. Extensive fires, followed by new shrub growth, allowed the population to recover and it seemed probable that the cycle would repeat itself. However by 1950 wolves had also reached the island and the two species have come gradually into an equilibrium, which is likely to prevent the previous population pattern of violent irruptions and collapses (Mech, 1972). Important in this balance is the fact that a healthy adult moose is well able to protect itself from a wolf pack and the greatest mortality occurs amongst calves, and aged or ailing individuals.

The accumulation of evidence to demonstrate the stabilising capabilities of wolves (Pimlott, 1967) has brought about a new attitude to these predators and it is now standard practice in much of Europe and North America to protect wolves already present in nature reserves and where possible to reintroduce them into areas where they have disappeared. In countries such as Britain where wolves have become completely extinct and are unlikely to be reintroduced, it is recognised that deer populations need to be culled artificially at a level commensurate with their natural rate of increase. In the Island of Rhum National Nature Reserve, for example, a sixth of the red deer (*Cervus elaphus*) population has to be removed each year to keep the population stable (Lowe, 1969).

Overcrowded elephant (*Loxodonta africana*) and hippopotamus (*Hippopotamus amphibius*) populations have caused habitat damage in a number of African National Parks (Laws, 1970). The hippos in Uganda's Rwenzori National Park have had to be culled for this reason, and similar measures have been taken against elephants in the Murchison Falls (Kabalega) National Park to check the destruction of wooded savanna. Here the damage is caused by the elephants' habit of stripping bark from the trees and uprooting them to feed on the roots and the foliage. The culling policy is not without its critics and in Kenya's Tsavo National Park, although the elephants were causing extensive damage, they were allowed to increase unchecked and are now suffering a population collapse as a result of food shortage. The important question is whether protecting elephants in parks in some way interferes with the normal processes of population regulation in the species. Certainly there is no parallel with the loss of wolves in the wolf/deer relationship; the elephant has no significant natural predators to lose.

Elephants have been shown to possess a mechanism based on reproductive checks for controlling population growth in crowded conditions (Laws, 1969). The mechanism is, however, very slow-acting and requires a number of years for its effect to become apparent. It is also known that in free-range conditions, elephants undertake continuous migratory movements which give habitats exploited for food, time to recover before the next invasion. These two factors may hold the key to the elephant problem. Most of the herds now confined in parks, even very extensive parks, are prevented from undertaking their previously extensive migrations. This is particularly well-illustrated by the Murchison Falls National Park (Fig. 2.9) but also applies to Tsavo (Myers, 1973). Deprived of this outlet the elephants' self-regulatory

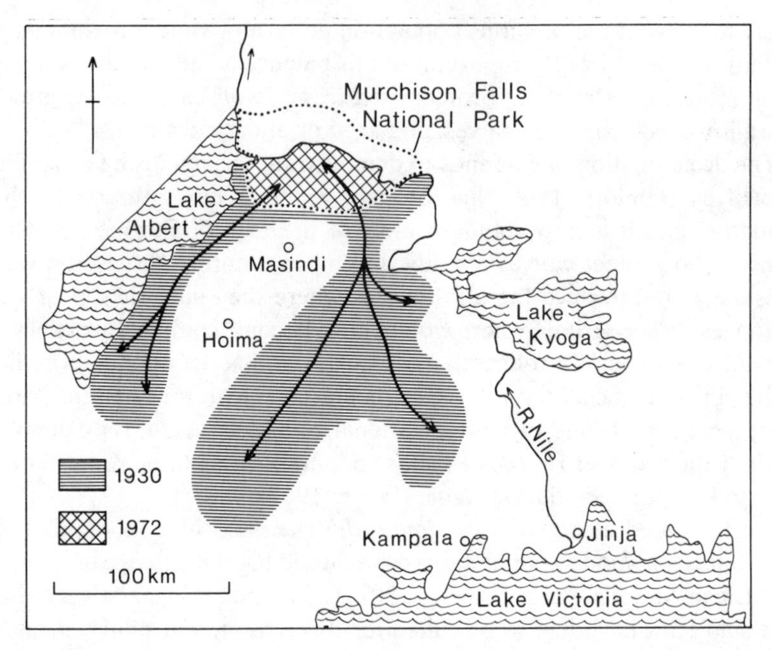


*Fig. 2.9* The curtailment of elephant migration routes associated with the establishment of the Murchison Falls (Kabalega) National Park (after Bere, 1975)

reproductive checks do not seem to be able to work quickly enough to prevent serious overcrowding and food shortage. On this basis artificial culling of park herds must be regarded as a necessary management technique.

## Recreation

In developed countries catering for outdoor recreation has become an increasingly important aspect of rural planning. The range of recreational activities is so wide as almost to defy classification. However, for the purposes of discussion it is useful to recognise the following major categories: (i) physical pursuits, typically involving tests of skill and endurance (e.g. climbing, caving, skiing), (ii) sight-seeing activities based principally on an aesthetic appreciation of the countryside and (iii) activities which have a specific involvement with wild life (e.g. natural history, hunting, shooting and fishing).

A number of problems arise in relation to recreational planning. Although it is the task of the planner to provide attractive recreational facilities, this can be self defeating because the pleasures of many recreational pursuits are diminished by overcrowding. Difficulties also arise from the frequent need to associate one recreation with another, or to combine recreational activities with other land (or water) uses. Some detailed examples of the contribution of ecology to these various problems of integration are described later in a case study of a National Park (p. 175).

Particular difficulties are raised by attempts to combine recreation and conservation interests, and some examples will be discussed below.

*Indirect recreational effects at conservation sites*

Many conflicts with biological conservation arise accidentally as the result of recreational activities which themselves have no intentional involvement with wildlife. Britain's upland and coastal regions provide numerous instances of this kind of interaction.

Much of the upland area has been transformed by sheep grazing from its natural state as woodland or species-rich heath, into moorland. Nevertheless, some near-natural communities still survive and amongst the most important are the fragments of arctic-alpine vegetation growing on steep slopes and cliffs which are inaccessible to sheep. In North Wales, the Snowdon National Nature Reserve has been established largely to protect these plant communities. However, as the reserve has many scenic attractions and includes Snowdon (1058 m), the highest mountain in Wales, it is also a natural focus of recreational activity. On a fine summer's day as many as 1500 people make their way to the summit via the various approach paths. A number of paths cross the arctic-alpine plant zones and with increasing recreational use there is a significant risk of trampling damage by visitors who stray from the paths. This has become particularly noticeable on the approach to the summit ridge from the eastern side, at the confused junction of the Miners Track and the Pyg Track, and on the col at Bwlch Coch which walkers use as an escape route from the exposed northern ridge of Crib Goch (Fig. 2.10).

To reduce the damage, the Park authorities in collaboration with the Nature Conservancy Council have instituted a series of measures aimed at checking diversions from existing paths. Short lengths of fencing have been erected to block short cuts and the existing paths have been made more attractive by smoothing boulder strewn sections, stabilising screes and reducing waterlogging. Interestingly, to find appropriate path-building methods it has been necessary to revert to the techniques used for road building before the introduction of cement and tar-macadam (Manasseh & partners, 1975).

Another conflict in Snowdonia involves rock climbing. Climbing tends to denude cliff of vegetation partly through simple wear and partly as a result of the practice of 'gardening' or stripping vegetation from climbs to make them safer. Although there are a large number of important climbing areas on

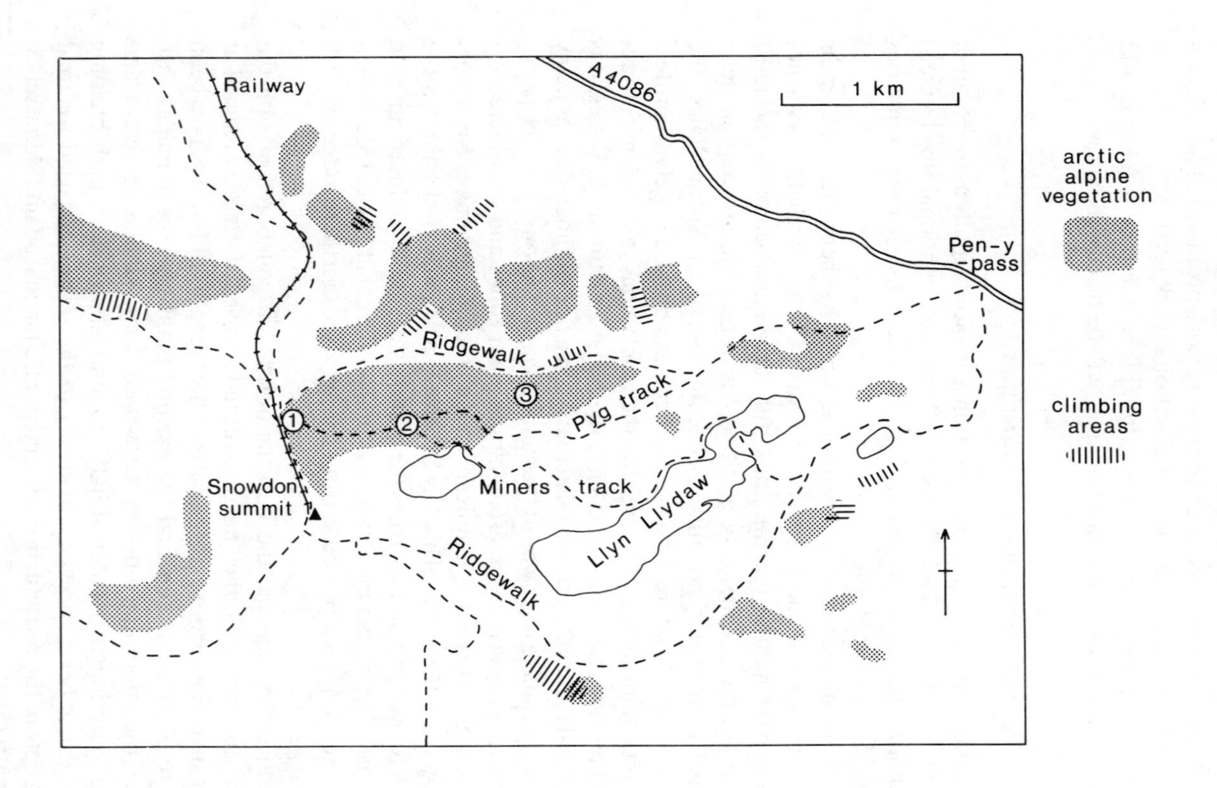


*Fig. 2.10* Potential conflicts between recreation and biological conservation interests in the Snowdon National Nature Reserve (partly after Manasseh & partners 1976). *1.* Eastern approach to summit ridge. *2.* Junction between Pyg Track and Miners Track. *3.* Bwlch Coch

Snowdon, few of them overlap with the arctic-alpine plant zones (Fig. 2.11). The separation has a geological basis. Whereas the interesting plant communities are associated mainly with the softer basic tuffs, these rocks are generally regarded as too soft and too easily fragmented for safe climbing. Climbers therefore concentrate almost exclusively on harder rhyolitic rocks and these support species-poor plant communities of no particular botanical interest. In one or two places where the approach to the base of the climbing pitches is across basic strata, there is a definite risk of trampling damage (Fig. 2.11). In these situations attempts have been made to acquaint the climbers with the problem involved and to persuade them to use a single approach route only.

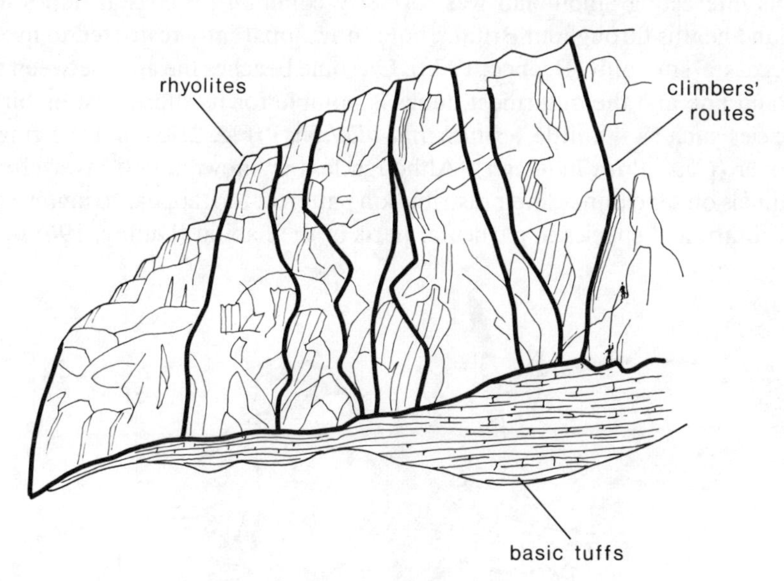


*Fig. 2.11* The arctic-alpine plant zone below Clogwyn-y-Ddysgl

*Problems at coastal sites.* Sandy beaches have been important as recreational areas ever since the fashion for sea-bathing developed in the 18th century. They are also often associated with interesting but very vulnerable biological systems. The wind-blown sand at the head of the beach becomes colonised by pioneer grasses, particularly marram grass (*Ammophila arenaria*), and later by a wider range of plants which completely cover the sand surface. In undisturbed conditions the different phases of this colonising process are laid out as a series of vegetation zones extending inland from the beach (Fig. 2.12).

These coastal systems support a range of animal species, many of them localised in particular parts of the habitat. Thus the natterjack toad (*Bufo*

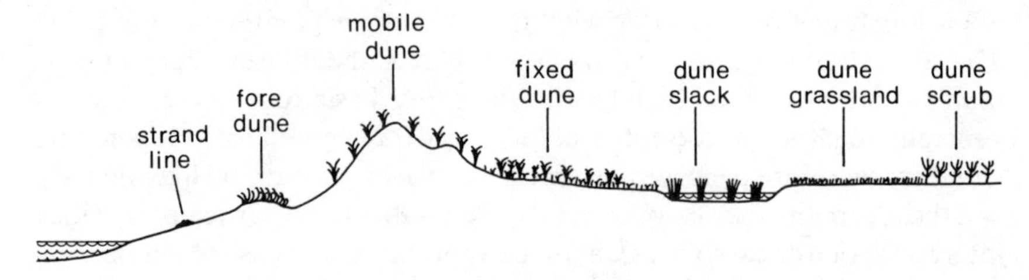


*Fig. 2.12* Typical vegetation zones on a sand dune system

*calamita*) is associated especially with dune pools and waterlogged hollows. This interesting amphibian was formerly common on coastal dunes and inland heaths throughout Britain, but is now apparently restricted to five or six coastal sites only (Beebee, 1973). On some beaches the area between the strand line and the outermost dunes is notable for its shore-nesting birds, species such as the little tern (*Sterna albifrons*) (Fig. 2.13) and the ringed plover (*Charadrius hiaticula*). Although less is known about invertebrate animals on sand dunes, their distribution patterns, too, appear to involve the localisation of species in particular parts of the habitat (Duffey, 1968).

*Fig. 2.13* Little tern (*Sterna albifrons*) (based on a drawing by R. Gillmor)

Recreational activities can disrupt these communities in a variety of ways. Most pressure is obviously concentrated on the beach and first line of dunes. Unfortunately these are often the most sensitive habitats. Nesting birds are readily disturbed by both people and dogs, whilst their eggs and young are liable to be trodden underfoot. On the fore dunes, trampling readily disrupts the first pioneer phase of plant colonisation. This can result in 'blow-outs' of sand which may in turn smother some of the established communities further inland. The fixed dunes and dune grassland can also be damaged by the regular passage of holiday-makers between car parks and the beach.

For areas of conservation importance, various simple management devices can be used to reduce the amount of damage. At Gibraltar Point Local Nature Reserve on the East Anglian coast holiday-makers moving from the car parks onto the beach are channelled onto two main paths, one of which is surfaced with old railway sleepers and bordered by a strong fence (Schofield, 1967). At Ainsdale National Nature Reserve on the North West coast, one of the few remaining refuges for the natterjack toad, visitors are strictly restricted to a limited number of specified paths. Blow-outs of sand are not easy to anticipate, but once they have occurred, various rehabilitation techniques can be used. These include spraying with latex, planting with rows of marram grass or combining grass-planting with the laying-down of a thatch of shrub or tree branches (Lloyd, 1970). The protection of shore-nesting birds is more difficult to arrange and usually needs to involve a combination of fencing, explanatory notices and active wardening (Chestney, 1971).

*Conflicts arising from wildlife-based recreation*

The prospect of diverting recreational impacts becomes more remote when the species being conserved itself represents a tourist attraction.

In contrast to Snowdonia's fairly inconspicuous arctic-alpine plants which few visitors even notice, the mountain flowers in some other parts of the world are sufficiently numerous and showy to represent a tourist attraction in their own right. In Colorado, thousands of tourists visit the Rocky Mountain National Park every summer to see the alpine flowers, and damage from trampling has become a major problem. Around parking areas the vegetation has been virtually destroyed and survives only as isolated tufts surrounded by a lower eroded gravel surface from which both vegetation and soil have disappeared. On such sites, even where protective fences have been maintained for four years, no recovery has taken place and it has been claimed that the natural vegetation will take a very long time, of the order of hundreds of years, to return, (Willard & Marr, 1970, 1971). Areas under less pressure stand a better chance of recovery and may show some significant improvement if protected for a few seasons. In an attempt to reduce the damage, efforts are being made to restrict visitors to specially-constructed paved walks.

*The Sequoia groves of California.* Although trampling damage frequently accompanies intense recreational use, there is a danger of concentrating on this factor to the exclusion of other possibilities. In the Sequoia, Kings Canyon and Yosemite National Parks of California the giant sequoia trees (*Sequoiadendron giganteum*), the last remnants of a once extensive mountain forest community, are a major tourist attraction. By the 1960s the Park authorities had come to the conclusion that soil compaction caused by trampling was limiting the growth of tree seedlings and might even be endangering the mature trees. Accordingly, in the Yosemite Park, seedling areas are fenced off and visitors were encouraged to view the sequoia groves at a distance from open buses rather than drive their cars or walk to the bases of the trees.

Subsequent studies suggest that the major factor suppressing the growth of young sequoias is probably not trampling at all, but the absence of fires (Vale, 1975). Sequoia seeds apparently develop best where minor fires have destroyed the surface layers of organic debris and have removed the early growth of competing tree species. Strict fire control in the Parks has prevented these optimum conditions from being realised and has been a major but unexpected factor in preventing regeneration. In keeping with this interpretation, the National Park Service has now started a programme of controlled burning in the Sequoia and Kings Canyon National Parks. In this instance trampling may actually have had a beneficial effect by producing the bare ground conditions suitable for sequoia germination.

*African game parks.* Amongst the tourist attractions based on animals, pride of place must go to the herds of game animals on the African plains. Although the combination of tourism and conservation in African parks makes sound economic sense, there are certainly situations where the presence of visitors is prejudicial to the welfare of the animals.

In the Nairobi National Park, lion cubs and zebra foals frequently become separated from their parents as a result of disturbance by tour minibuses. Quite often it seems that the disturbed species reacts to human intruders with a piece of stereotyped behaviour which has evolved for some other purpose. In the Ngorongoro Crater, Thomson's gazelles (*Gazella thomsoni*) react to visitors' cars in the same way as they react to competitors and predators. The female leaves the breeding territory well in advance of the male who delays his departure until the intruder has come closer. With the increasing number of visitors' cars, the females are constantly moving out of the breeding territory and being separated from the males. This is probably reducing the breeding success of the species (Walther, 1969).

Similarly the presence of tourists has been shown to be detrimental to the breeding of crocodiles, in this case by rendering the eggs and young more vulnerable to predators.

In the Murchison Falls National Park, Nile crocodiles (*Crocodilus niloticus*) are a major tourist attraction and regular boat trips are organised to view the river-bank breeding areas. Female crocodiles actively protect their eggs, which are buried in the sand during a three month incubation period, and also defend their young. The presence of tourist boats has the effect of driving females from the bank into the water and this allows predators, particularly monitor lizards and olive baboons to move in and take buried eggs or young (Cott, 1969). Monitor lizards can be seen to devour a dozen eggs in 75 minutes and baboons often depart carrying three or four eggs apiece. In consequence losses to predators are significantly greater at localities visited by tourists than those which are left undisturbed (Table 2.6).

*Table 2.6* The effect of tourists on the loss of crocodile nests to predators (from Cott, 1969).

| *Breeding ground* | *Number of nests* | *Nests destroyed by predators* |
|---|---|---|
| | *Tourist sites* | |
| Sand-river 4 | 13 | 13 (100%) |
| Namsika | 10 | 10 (100%) |
| Bee-eater slope | 13 | 7 (54%) |
| | *Sites not visited by tourists* | |
| Falls Bay | 36 | 17 (47%) |
| Frog Bay | 7 | 2 (29%) |
| Sand-river 2 | 22 | 4 (18%) |
| Paraa South | 7 | 0 (0%) |

*Coastal breeding colonies.* In many parts of the world breeding colonies of seabirds and seals have also become major tourist attractions. Breeding seals are particularly sensitive to disturbance and the survival of some colonies could be threatened if tourist pressures become more intense. With seabird colonies, as with crocodiles, visitors have the effect of increasing losses to predators. At Punto Tombo, a Nature Reserve in northern Patagonia it has been shown that the entry of visitors into the breeding colonies of king shags (*Phalacrocorax albiventer*) (Fig. 2.14) and magellanic penguins (*Spheriscus magellanicus*) increases the loss of eggs to predatory gulls (Kury & Gochfield, 1975). As a visitor moves into the shag colony the parent birds in the first ten rows or so leave their nests and the gulls, which are constantly patrolling the colony edge, move in to steal the eggs. In other colonies young birds are also taken. The gulls seem to recognise the fact that the presence of human visitors creates the opportunity to attack the nests.

In most of these situations, the problems can be reduced by arranging for a greater measure of separation between the animals and the visitors. In the Ngorongoro Crater Conservation Area disturbance to the gazelles would be reduced by limiting visitors' cars to specified routes. In the Murchison Falls

*Fig. 2.14* King shag (*Phalocrocorax albiventer*)

National Park significant disturbance to crocodiles would be prevented by excluding visitors during the four-month breeding season. Obviously as the weight of restrictions increases, so the level of recreational enjoyment is likely to decline and it has to be recognized that there are some situations where conservation and recreation interests cannot coexist. More options are available when the natural history interests of visitors are less specific. It may then be possible to use 'show-piece' exhibits to divert the attention of visitors from sensitive areas. At Scolt Head Island Nature Reserve, visitors are shown terns' nests containing eggs taken from deserted nests, as a substitute for viewing the disturbance-sensitive breeding colonies (Chestney, 1971). Nature trails, guided walks and observation points can also be used for diversionary purposes. The Bayerischer Wald National Park in Western Germany, which has been established to protect and demonstrate the plant and animal communities of a European forest, provides an interesting example of this strategy translated into spatial terms (Fig. 2.15). It is arranged in three zones of decreasing accessibility. In the first zone, adjacent to the car parks and access points, large enclosures have been set up to exhibit forest animals such as wolves, beavers and European bison in conditions where they can be easily observed and photographed. In the next zone access is still virtually unrestricted but the movement of visitors is channelled using nature trails. The innermost zone has been devised as a refuge where the woodland community can exist in near-natural conditions. This zone is available for scientific research studies but other visitors are restricted to specified paths.

*Shooting as a recreation.* Recreational shooting is another major wildlife-based activity which needs to be considered in relation to conservation. There is an increasing tendency to regard the two interests as compatible or even to represent shooting as an aid to conservation. As we have seen, there

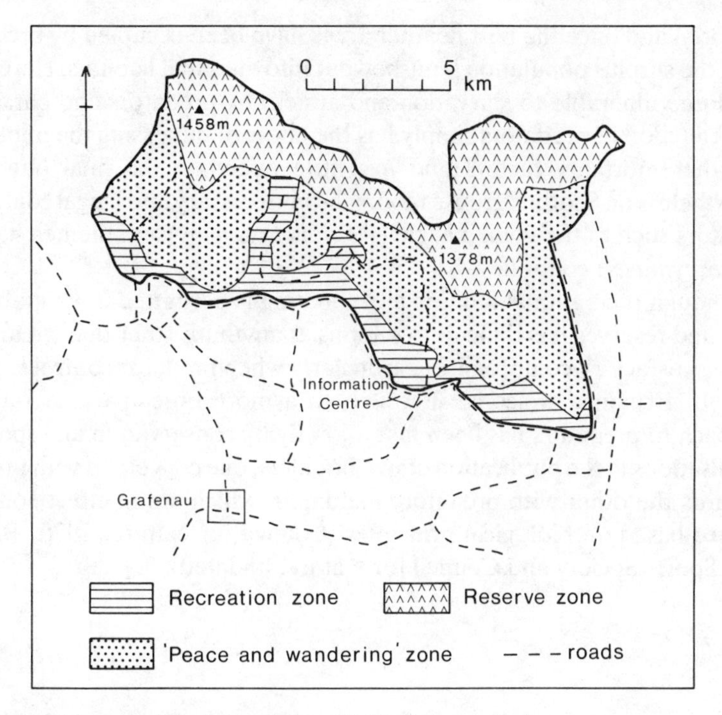


*Fig. 2.15* Zoning of the Bayerischer Wald National Park

are occasions where animals in nature reserves or parks need to be culled in their own interests and to prevent damage to the habitat. It has been argued that there is no difference between this operation and the exploitation of a surplus population for sporting purposes.

In fact, the management implications of the two approaches are quite different. The conservationist's ideal must be to maintain balanced, self-regulating natural communities with the minimum of human intervention. By contrast the sportsman seeks not only to insert himself into the system as a predator, but also to manipulate conditions to produce maximum numbers of his chosen prey. This is one fundamental impediment to the fusion of shooting and conservation interests. The other is concerned with attitudes to natural predators; most sportsmen see themselves as being in competition with natural predators and 'vermin control' has always been regarded as one of the gamekeeper's functions. This attitude to predators persists even where it is based on manifestly unsound premises. A most thorough investigation of red grouse (*Lagopus lagopus*) populations in Scotland has shown that activities of predators have no significant effect on the numbers of grouse available to the sportsman when the shooting season opens in August (Jenkins *et al.*, 1964, 1967). Essentially this is because grouse are highly territorial and once the best heather areas have been occupied by breeding birds, the surplus population is pushed out into marginal habitats. Here they are more vulnerable to starvation and attacks by predators and parasites. The reduction of predators simply has the effect of increasing the impact of the other mortality factors and has little effect on the final outcome. Nevertheless on Scottish grouse moors the unnecessary and illegal control of predators such as the golden eagle (*Aquila chrysaetos*) and the hen harrier (*Circus cyaneus*) continues.

Although recreational hunting and shooting is permitted in a number of parks and reserves it is difficult to see this as anything other than an uneasy and unsatisfactory compromise, particularly when predator control is practised. In Britain a welcome first step to a more objective and rational approach to predators has been taken jointly by conservation and sporting organisations in the publication of two booklets, one concerned with predatory birds the other with predatory mammals, which are in effect codes of practice based on biological principles (Council for Nature, 1973; British Field Sports Society and Council for Nature, undated).

# 3 Urban development

By their very nature, towns have to accommodate a much wider range of different functions than rural areas. There must be provision for houses, workplaces, open spaces for recreation, transport systems, and services to supply water, power, food and remove wastes. The aim of planning is to harmonise these functions and produce an urban environment which is pleasant and healthy whilst remaining efficient. For a variety of reasons these objectives are not always achieved. Sometimes one element develops disproportionately at the expense of others. Many residential areas in towns are subjected to intrusive noise levels from roads and airports which have been expanded to serve wider regional functions. These problems and others associated with transport systems are considered in detail later (p. 121).

Probably the most widespread source of environmental problems in towns arises from inadequacies in the disposal of waste materials of various types. A typical modern city of a million people produces in one day something like 500 000 tons of sewage, 2000 tons of domestic refuse and 950 tons of gaseous and particulate matter discharged into the atmosphere (Wolman, 1965). The capacity of these wastes to cause trouble is continually under-estimated. Some wastes create problems because they are injurious to life and health, others because they can sustain a whole array of troublesome species including pathogens, parasites, disease-carriers and scavengers. It will be seen that the ecology of towns is largely the story of difficulties arising from inadequate waste disposal systems.

## Air movements and atmospheric pollution

Town atmospheres have a considerable capacity for cleansing themselves of pollution because of the frequent occurrence of rising currents of warm air which disperse wastes upwards and draw in clean air from the surrounding countryside. The circulation pattern results from fuel-burning and the high thermal capacity of building materials in towns, which makes them warmer than rural areas. Unfortunately, this cleansing mechanism is sometimes brought to a halt by layers of static air which move in above the town and trap pollutants like the lid on a saucepan. This occurs when a layer of cold air becomes sandwiched between the ground and a higher layer of warm air. It is a reversal of the normal situation where the air layers nearest the ground are warmest and those at increasing altitudes progressively colder, hence the term 'inversion' (Fig. 3.1).

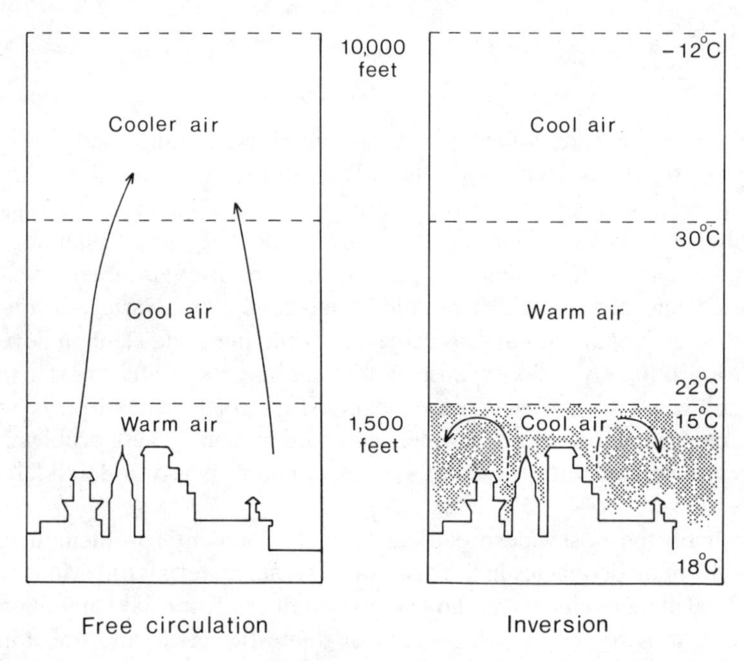


*Fig. 3.1* Comparison of free circulation and inversion conditions over a city (data for New York from Benarde, 1973)

Inversions can occur through a variety of causes. One of the simplest situations is where a town is located in a valley and air which has been cooled on the hill tops, thereby becoming more dense, flows down the valley slopes and becomes trapped on the valley floor beneath warmer layers. It was a situation of this type which gave rise to the serious pollution incidents in the Meuse Valley in Belgium in 1930 and in the town of Donora, Pennsylvania, in 1948, causing respiratory illnesses in thousands of people and a number of deaths. Not all inversions are associated with valleys, however, they can also form in relatively open country during clear periods in winter when heat is rapidly lost from the air layers near the ground. Neither London nor New York are in typical valley situations, yet both cities have suffered seriously from pollution associated with inversions. The best documented were the

London 'smogs' of 1952, 1956, and 1962, which were responsible for widespread respiratory illnesses and nearly 6000 deaths (Royal College of Physicians, 1970).

The origin of the polluting material varies from one situation to another. In London the major pollutants were sulphur dioxide, particulate matter and droplets of tar derived from burning soft coal on open fires. In the Meuse Valley and at Donora, pollutants from domestic sources were supplemented by emissions from chemical plants and metal smelters.

In a few situations exhaust emissions from motor vehicles are the dominant air pollutants. The well-known photochemical smog of Los Angeles is largely the result of a series of interactions involving hydrocarbons and oxides of nitrogen from car exhausts (Fig. 3.2). In the presence of strong

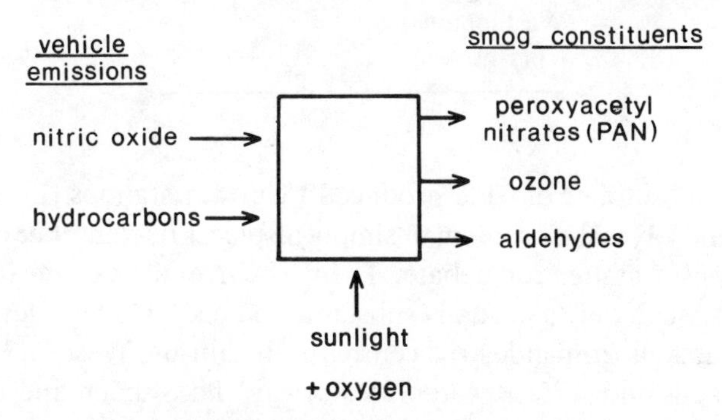


*Fig. 3.2* Initial reactants and end-products in photochemical smog formation

sunlight a substance, peroxyacetyl nitrate or 'PAN', is formed which can cause serious eye irritations and crop damage (Haagen-Smit, 1968). Inversions are involved here too, because the hills surrounding the city on three sides tend to trap cold air masses on the valley floor.

Although photochemical smog has come to be identified with Los Angeles, any city with a large number of motor vehicles, a sunny climate and a susceptibility to inversions can experience the same problems, although usually on a smaller scale. The oxidants associated with photochemical smog have been detected in appreciable concentrations in Genoa, Bombay, Sydney and a range of North American cities (Table 3.1).

## *Remedial measures*

There are a number of possible ways of combating urban atmospheric pollution. With large industrial installations it is often feasible to build chimneys sufficiently tall to penetrate the inversion layer and discharge their wastes beyond it. This strategy has been widely adopted in Britain for

*Table 3.1* Number of days per annum on which the oxidant level exceeded 0.15 p.p.m. for 1 hour. At 0.15 p.p.m. approximately 50 per cent of the population experience eye irritation (Patterson and Henein, 1972)

| | |
|---|---|
| Los Angeles | 115 |
| San Diego | 19 |
| Denver | 13 |
| Philadelphia | 8 |
| St. Louis | 6 |
| San Francisco | 4 |
| Washington | 3 |
| Chicago | 1 |

dispersing the sulphur dioxide produced by power stations (Lucas, 1974). Whether this solves the problem or simply displaces it to another geographical locality is a matter for debate. There is currently a claim from Scandinavia, although only a partially substantiated one, that high-level sulphur dioxide emissions from industrial centres in Britain and Western Europe are acidifying soils and surface waters in Norway and Sweden and interfering with forest production and fisheries (Department of the Environment, 1976a).

The scattered pollution sources represented by domestic chimneys and small industrial premises require a different approach and here Clean Air legislation has been effective in checking the discharges at source by stimulating a transfer to smokeless fuels. The control of motor vehicle emissions which is probably necessary only in regions susceptible to photochemical smog is being approached through engine design and exhaust system modifications (p. 132).

Finally, it must be admitted that some industries even when employing the latest technological advances in waste treatment still continue to pollute the urban atmosphere. A dust-arrestor which is 99% efficient may still let through 500 kg of particulate matter every hour. Yet there are often compelling arguments, based on maintaining employment opportunities, for associating industries with towns. A simple solution is to site the offending industry downwind of sensitive residential and city centre areas. McHarg (1969) argues that planners should go further than this and recognise the scope for 'airshed' planning. He cites the case of Philadelphia which at the time of his study (1963) had a recognisable pollution core during inversion conditions. This pollution core was capable of being swept clear, albeit very slowly, by the weak winds prevailing during inversions, which came predominantly from the west, north-west, and north-east. From this knowledge of wind directions supplemented by information on typical windspeeds it was possible to define on the map of the city three airzones or airsheds which were capable of performing this cleansing function (Fig. 3.3). It was proposed that these zones should be kept uncontaminated by excluding new polluting industries from them.

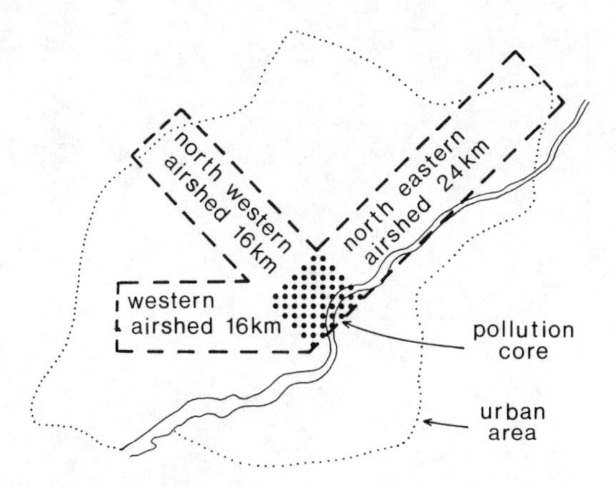


*Fig. 3.3* Airsheds for Philadelphia (after McHarg, 1969)

## Organisms associated with human wastes

Many of the diseases which afflict man produce infective stages in his faeces and even when faecal material is dispersed in soil or water these infective stages remain viable, some for a period of days, others for weeks and a few for months (Table 3.2). Urban sanitation systems are intended to check the spread of infections by removing contaminated material from human contact until the danger period is passed. For a variety of reasons this aim is not always achieved and many rapidly expanding towns in the tropics are characterized by a high incidence of enteric diseases resulting largely from inadequate waste disposal and water supply systems.

### *Parasites in the soil*

One problem appears characteristically in situations where a proportion of a town's population fails to utilise such sanitary facilities as are available and defaecates at random during the course of a normal day's activity. This practice can cause widespread contamination of urban soils and increases

*Table 3.2* Diseases associated with the faecal contamination of soil and water in urban areas

| Disease | Causative organism | | Principal infection routes | | | Survival time in water (W) and soil (S) | | | |
|---|---|---|---|---|---|---|---|---|---|
| | | | *Food* | *Water* | *Soil* | | *Days* | *Weeks* | *Months* |
| Hookworm infection | *Ancylostoma duodenale* or *Necator americanus* | nematode worm | | | × | (S) | | × | |
| Threadworm infection | *Strongyloides stercoralis* | nematode worm | | | × | (S) | | × | |
| Large intestinal roundworm infection | *Ascaris lumbricoides* | nematode worm | × | | × | (S) | | | × |
| Whipworm infection | *Trichuris trichiura* | nematode worm | × | | × | (S) | | × | |
| Amoebic dysentery | *Entamoeba histolytica* | protozoan | × | × | × | (W), (S) | | × | |
| Cholera | *Vibrio cholerae* | bacteria | × | × | | (W) | | × | |
| Typhoid fever | *Salmonella typhi* | bacteria | × | × | | (W) | × | | |
| Bacillary dysentery | *Shigella* spp. | bacteria | × | × | | (W) | × | | |
| Poliomyelitis | Polio virus | virus | × | × | | (W) | × | | |

the infection risk from a number of important parasites which live parasitically in man's alimentary tract. These include some nematode worms, *Ancylostoma*, *Necator* and *Stronglyoides* (Fig. 3.4*d* and *f*) which gain access to the body when their larval stages penetrate the skin. This happens most frequently when bare feet make contact with faecally contaminated soil. In a second group which includes *Ascaris* and *Trichuris* (Fig. 3.4*c* and *e*) the

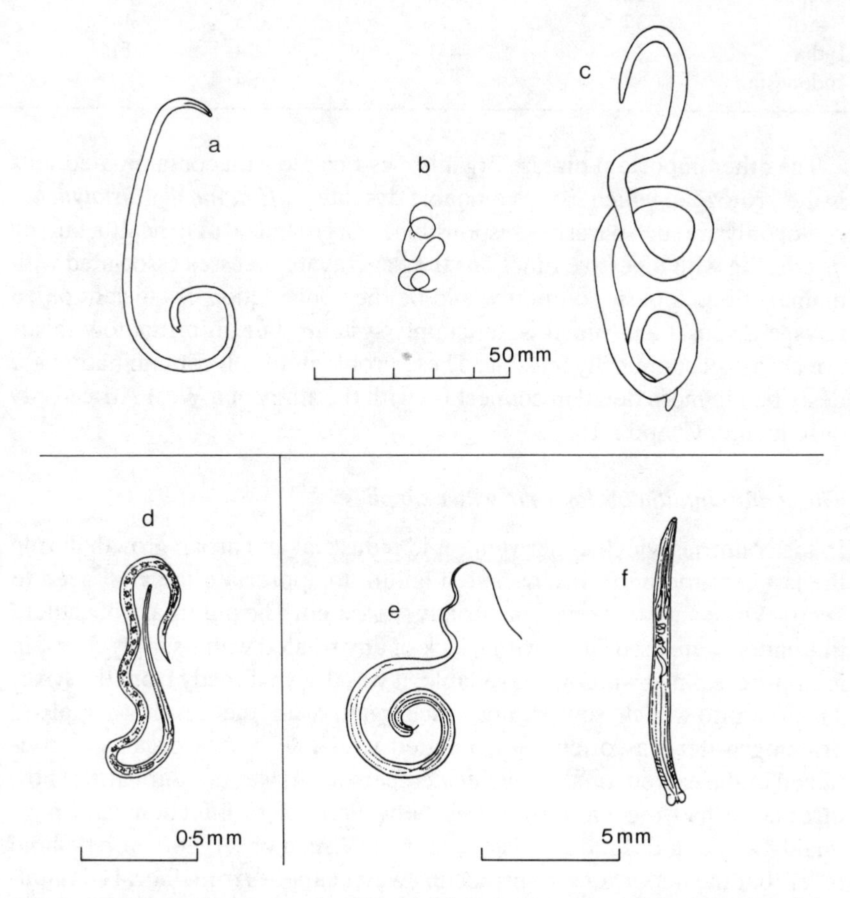


*Fig. 3.4* Nematode worms of public health importance in urban situations: (*a*) *Toxocara canis*; (*b*) *Wuchereria bancrofti*; (*c*) *Ascaris lumbricoides*; (*d*) *Strongyloides stercoralis*; (*e*) *Trichuris trichiura*; (*f*) *Ancylostoma duodenale*

route of infection is via the mouth, the eggs of the parasite being picked up directly from contaminated soils or from garden produce grown on plots where 'nightsoil' has been used as a fertiliser. Table 3.3 shows that these nematode parasites often have a high incidence in tropical countries.

*Table 3.3* Percentage infection by soil-transmitted nematodes in surveys from various countries (World Health Organisation, 1964)

| | *Necator/ Ancyclostoma* | *Strongyloides* | *Ascaris* | *Trichuris* |
|---|---|---|---|---|
| Puerto Rico | 27 | 5 | 31 | 80 |
| Mexico | 26 | 3 | 33 | 28 |
| Brazil | 37 | – | 75 | 44 |
| India | 20 | 1 | 14 | 11 |
| Indonesia | 89 | – | 64 | 87 |

The other important disease organism associated with contaminated soils is the protozoan which causes amoebic dysentery, *Entamoeba histolytica.*

Not only are such parasites responsible for a great deal of ill health, but, by interfering with digestive functions they aggravate illnesses associated with malnutrition. A basic solution would be the construction of efficient piped sewage disposal and piped water supply systems, but unfortunately this is not always economically feasible. These problems of soil contamination are described in more detail in connection with the study of a West African city described in Chapter 10.

*The contamination of domestic water supplies*

In the countries which underwent rapid industrial and urban growth during the last century there was, at first, a failure to appreciate the vital need to keep sewage and water systems strictly separated. The outbreaks of cholera in London were associated with a lack of any treated water supply. Even in areas where piped water was available, it was drawn directly from the River Thames into which foul drains discharged, with the result that piped drinking water was often contaminated with the disease organisms contained in the excreta of already infected persons. It was not until after Snow in London had recognised the link between cholera infections and poor quality water, and Budd in Bristol had made similar observations on typhoid fever that the importance of protecting water supplies from faecal contamination was realised. Even with this knowledge gaps between principle and practice still remain. Many water treatment and supply systems in developing countries have been designed along the right lines but have been overwhelmed by the increasing demands of growing urban populations, and here the distribution of contaminated water to consumers continues.

Even in highly developed countries with excellent piped water facilities contamination can occur in exceptional circumstances. The Swiss holiday resort of Zermatt suffered a typhoid epidemic in 1963 which involved 437 cases and 3 deaths and was apparently attributable to the leakage of sewage into a water chlorination tank (Bernarde, 1973). In Szeged, a Hungarian city of some 100 000 people, an outbreak of enteric diseases in 1955 was traced to the entry of sewage into the water supply system (Fig. 3.5). Contamination followed the breakdown of a sewage pumping station at a time when three days torrential rain had caused flooding. Altogether the resultant epidemic of enteritis, typhoid fever and hepatitis affected 30 000 people, almost 30 per cent of the city's inhabitants.

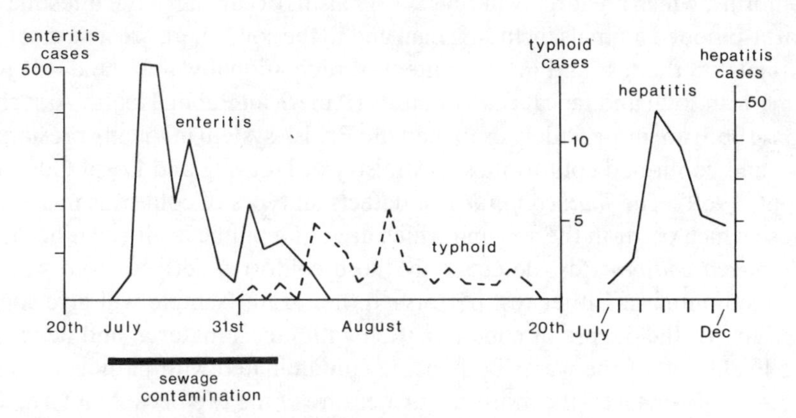


*Fig. 3.5* Disease outbreaks in the Hungarian City of Szeged following sewage contamination of the water supply system in 1955 (after Bakács, 1972)

The narrow safety margin involved in urban water supplies becomes even more evident in times of natural disasters such as earthquakes or floods, or during civil disturbances. In these circumstances particularly stringent measures are necessary to protect the integrity of the supply and counter the spread of disease (Assar, 1971).

## *The contamination of recreational waters*

Even in those communities where the first priority of protecting water supplies from faecal contamination is achieved, there are great variations in the methods of sewage disposal used. At one extreme the sewage may receive no treatment at all and be discharged directly into rivers or coastal waters. For many urban authorities this is a convenient way of transferring their waste problems to another area. At the other extreme, the treatment may produce an effluent which is virtually free not only of solid materials but also of sewage-derived mineral salts such as phosphates and nitrates.

It has been questioned in recent years whether the less complete methods of sewage treatment produce an effluent which is sufficiently free of pathogenic organisms to be discharged into waters which are used for public recreation. A number of the pathogens derived from the alimentary canal of infected persons can survive for some days or even weeks in water (Table 3.2) and therefore could represent a health hazard for bathers.

For this reason, in many parts of the world, standards have been devised to express the degree of cleanliness required for a recreational water. These standards have been based on bacteriological tests commonly used to assess the purity of drinking water. They use the same group of bacteria, the coliforms, which are largely harmless organisms occurring in the intestine of warm-blooded animals including man and in the soil. There are a number of variants on the test and in the context of recreational water standards the American total and faecal coliform tests (American Public Health Association, 1965) are more widely used than the British system involving presumptive and confirmed coliform tests (Ministry of Housing and Local Government, 1969). The *total coliform* test detects all types of coliforms including those which occur in the soil and which are of no public health significance; the *faecal coliform* test detects only those coliforms derived from warm-blooded animals. Either test performed on a water sample will give some measure of the degree of contamination with faecal material and hence of the likelihood of the water body being contaminated with pathogens. The faecal coliform test is the more useful measure of the two because it virtually eliminates the extraneous influence of soil coliforms and other bacteria.

There are remarkable variations in the standards adopted by various authorities for recreational waters (Table 3.4). Recommended total coliform standards range from 50 to 800 000/100 ml and faecal coliform standards from 70 to 5000/100 ml. Some countries, including Britain, have been reluctant to adopt any standards. The British attitude was largely influenced by the pronouncements of a committee set up to consider possible health risks from bathing in sewage-polluted sea water (Medical Research Council, 1959). The committee took the view that 'a serious health risk is probably not incurred unless the water is so fouled as to be aesthetically revolting'. Water contaminated to this degree would be likely to have coliform counts exceeding all the standards quoted in Table 3.4. It is now likely that Britain will adopt the standards proposed by the European Economic Community.

All this still leaves open the question of whether bathing in sewage-contaminated water, either salt or fresh, can represent a health risk and if it does, what standards should be applied to minimise the risk. The fact that many people report minor illnesses after bathing at coastal resorts is not in itself evidence that sewage-pollution is responsible. Bathing and diving, even in the cleanest water, can cause micro-organisms already present in the nose, throat or ears, to move to new sites and so set up local infections. An

*Table 3.4* Bacteriological standards for bathing waters

*Total coliform standards*

| *Upper limit for total coliforms/100 ml* | *Country/State, etc.* |
|---|---|
| 800 000 | U.S. Environmental Protection Agency |
| *10 000 | European Economic Community |
| 5 000 | Illinois |
| 2 400 | Virginia, New York |
| 1 600 | Louisiana |
| 1 000 | U.S.S.R., U.S.A - 23 States |
| 500 | Vermont |
| 240 | Idaho, New Hampshire |
| 200 | Arkansas |
| 50 | Utah |

*Faecal coliform standards*

| *Upper limit for faecal coliforms/100 ml* | | *Country/State, etc.* |
|---|---|---|
| *5 000 <20°C | Seawater | European Economic Community |
| *2 000 >20°C | Seawater | European Economic Community |
| *2 000 | Freshwater | European Economic Community |
| 1 000 | | Alabama, Georgia, Mississippi, Tennessee |
| 240 | | Maryland |
| 200 | | U.S.A. - 12 States |
| 100 | | Colorado, Michigan |
| 70 | | Virgin Islands |

* 95 per cent of samples must be equal or lower than value specified.

interesting survey recently carried out by the consumer associations of France and Belgium concluded on the basis of 9000 replies that the chances of becoming ill after a coastal holiday were approximately doubled if the holiday-maker bathed in the sea. Some of the disorders, such as ear infections and sinusitis, which showed an increased incidence in bathers could certainly have been produced by the bathing *per se*, irrespective of the cleanliness of the water. Other infections which showed an increase in bathers as compared with non-bathers were more likely to have been linked with sewage pollution. For example, there were 23 cases of hepatitis and 252 cases of enteritis amongst bathers, whereas the comparable figures for non-bathers were nil and 32 respectively. In most areas, health authorities have cases in their records where there is circumstantial evidence to associate illness with sewage pollution. For example, in the Cardiff region a typhoid case in 1955 and a paratyphoid case in 1956 each involved children who swam regularly in a river in the vicinity of a sewage outfall.

A major difficulty in arriving at realistic standards for recreational waters is the lack of quantitative information relating the incidence of illness to the degree of sewage pollution. One exception is the study by Stevenson (1953) which showed that there was a direct relationship between the incidence of gastrointestinal disorders and the level of sewage pollution. On Lake Michigan the maximum health effect was recorded when the total coliform counts reached an average of 2300/100 ml; the comparable value for the Ohio River was 2700/100 ml. This would suggest that at least some of the standards listed in Table 3.4 are too stringent and could lead to the public being denied access to safe sites. By the same token some standards would seem to be too lax, possibly encouraging the use of sites which are unsafe.

The nature of the coliform test itself adds to the difficulty of arriving at precise standards. One problem is that even the faecal coliform test includes in its result, organisms derived from animal faeces which although not irrelevant to human health are generally less important than those from human excreta. A possible way round this difficulty is to combine the faecal coliform test with one for faecal streptococci. Faecal coliform bacteria are more numerous than faecal streptococci in domestic sewage of predominantly human origin, whereas the reverse is true for the faeces of other animals. This means that the ratio of faecal coliforms to faecal streptococci can be used to give an indication of the source of pollution (Table 3.5). In a number of situations examined in South Wales there was good agreement between the predictions made on the basis of the ratio and the directly observable sources of contamination (Table 3.6).

Another difficulty associated with the coliform test is that there are some situations where pathogens can be present in a water, even though the coliform count is low. This can come about, for example, where a typhoid

*Table 3.5* The use of the faecal coliform/faecal streptococci ratio as an indication of pollution sources (Geldreich, 1967; Geldreich and Kenner, 1969)

| *Faecal coliform/ Faecal streptococci ratio* | *Likely source of pollution* |
|---|---|
| >4.0 | Human |
| 4·0–2·0 | Mixed pollution, but predominantly human |
| 2·0–1·0 | Uncertain |
| 1·0–0·7 | Mixed pollution, but predominantly animal |
| <0·7 | Animal |

*Table 3.6* Some worked examples from South Wales, comparing the conclusions from the faecal coliform/faecal streptococci ratios with direct observations on likely sources of contamination (Data provided by G. H. Tarbutt, University College, Cardiff)

| | *Faecal coliform/ Faecal streptococci ratio* | *Conclusion from ratio* | *Observable sources of contamination* |
|---|---|---|---|
| Urban sewer | 4·4 | Human | Human sewage |
| River in coalfield valley town | 1·24 | Uncertain | Human sewage, drainage from sheep pasture |
| Streams in dairy farming area | 0·82 | Predominantly animal | Washings from animal houses; septic tank drainage |
| Urban gutter in coalfield valley town | 0·46 | Animal | Dog faeces; sheep faeces |

carrier is contributing to a small-scale sewage discharge to a river. The only way to detect this kind of anomaly is to test directly for pathogens such as *Salmonella typhi*. This is feasible but is usually considered too laborious to adopt as a routine procedure.

The survival rate of enteric bacteria is longer in freshwater than in salt water, raising the question of whether the recreational standards for the two habitats should be different. The E.E.C. standards are the first to take some account of this factor.

In spite of these various complications, there would seem to be a *prima facie* case for believing that sewage-polluted recreational waters can represent a health hazard and that such hazards are most easily assessed on the basis of coliform tests. This being so, a more incisive approach to the formulation of standards is urgently needed.

## *The contamination of shellfish*

A more obvious health hazard arising from the coastal discharge of sewage is that associated with the contamination of edible shellfish. Bivalve molluscs such as clams, mussels and oysters, living in sewage-contaminated waters can accumulate organisms of faecal origin in the course of their normal filter-feeding activities. This accumulation makes the animals unfit for human consumption in the raw state and there have been a number of outbreaks of typhoid, paratyphoid and infective viral hepatitis in Europe and North America attributable to the consumption of uncooked shellfish. The problem can be solved by subjecting the animals to a cleaning process before marketing. This involves placing them for a few days in uncontaminated coastal areas or in tanks containing sea water which has been sterilised with chlorine, ozone or ultraviolet radiation (World Health Organisation, 1974b).

*Sewage pollution and disease-carrying insects*

Quite apart from its significance as a course of enteric parasites and pathogens, faecal material can affect disease patterns by creating suitable breeding places for mosquitoes. There are a number of disease-carrying species which thrive in the sewage-enriched waters of towns.

Two such mosquitoes in the United States are *Culex pipiens quinque fasciatus* and *Culex tarsalis* which are important vectors of the Western and St. Louis encephalitis viruses. *C.p. quinquefasciatus* breeds all year around in the sewage and storm water disposal systems of some American cities and *C. tarsalis* is associated particularly with shallow sewage lagoons.

In the tropics and subtropics the important mosquito of sewage-enriched sites is *Culex pipiens fatigans*. This is certainly a near relative of *C.p. quinque fasciatus* and may even be the same species. It is important as the vector of the filarid worm *Wuchereria bancrofti* (Fig. 3.4*b*) which causes filariasis, and in extreme cases elephantiasis (Wilcocks & Manson-Bahr, 1972). Its favoured breeding sites include septic tanks, soakaways, latrine sumps and polluted open drains (Fig. 3.6). Urbanisation is often accompanied by dramatic increases in *C.p. fatigans* populations. For example in Kaduna in northern Nigeria a survey in 1942 failed to reveal any *C.p. fatigans* whereas by 1960 they were being recorded in densities of up to 760 per room per night (Mattingly, 1969). In the Indian city of Hyderabad, a mosquito survey in 1943 revealed no *C.p. fatigans*. In 1962, by which time the human population of the City had doubled, the mosquito was breeding profusely and filarial transmission was occurring. In Bangalore in Mysore State, a city previously renowned for its freedom from mosquitoes, the spread of filarial infections by *C.p. fatigans* has become increasingly important with the expansion of the city and a decline in sanitation standards (Singh, 1967).

This particular mosquito is especially difficult to control. It is virtually impossible to eliminate organically-polluted surface waters from towns and spraying the breeding sites with insecticides may not always be effective because in some areas the species has developed resistance, sometimes of a high level, to DDT, dieldrin and other insecticides (Tadano & Brown, 1966).

The transmission of filarid worms by *C.p. fatigans* is rightly regarded as potentially the most serious of all mosquito-based problems in towns.

Whereas protection against other mosquito-borne diseases such as yellow fever and malaria can be achieved by immunisation and prophylactic drugs respectively, there is still no completely effective method of preventing or treating filarial infections.

Mercifully, the pollution of urban waters makes them less attractive as breeding places for some mosquitoes. This is the case in Africa with important malaria carriers such as *Anopheles gambiae* and *Anopheles*

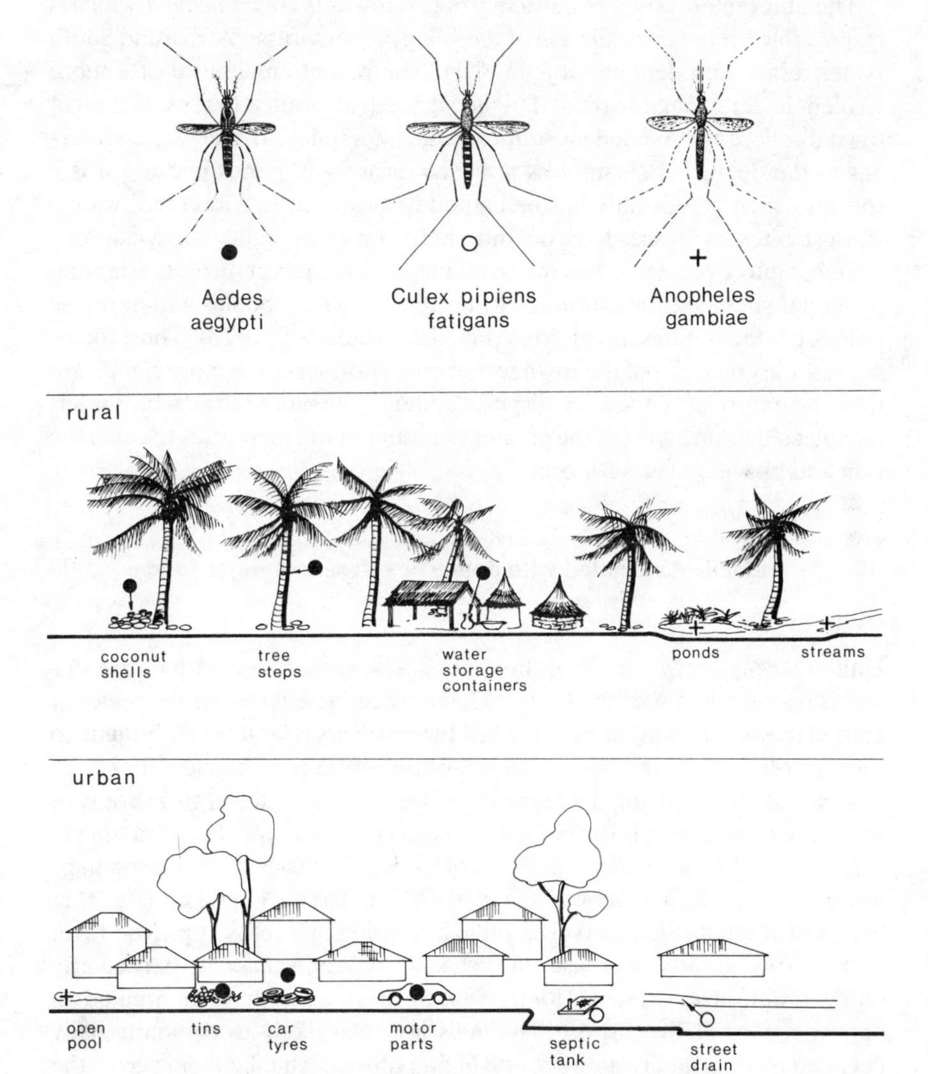


*Fig. 3.6* The effect of urbanisation on the availability of mosquito breeding sites in West Africa

*funestus*. They have a preference for clean water habitats such as ponds and small streams in open situations. Whilst these are readily available in rural areas, they become increasingly difficult to find in towns. The same is true of Asia where the malaria mosquitoes *Anopheles culicifacies and Anopheles minimus* decline as clean water habitats disappear (Surtees, 1971). Consequently, urbanisation is frequently accompanied by an increase in *C.p. fatigans* but a simultaneous decrease in malaria mosquitoes (Fig. 3.6).

The other important mosquito in tropical towns is *Aedes aegypti* which is responsible for the transmission of the yellow fever virus in Africa and South America and the dengue virus in Asia. The recent emergence of a more virulent haemorrhagic form of dengue has been attributed to sensitisation of town dwellers by repeated mosquito-borne infections. *A. aegypti* is indifferent to the degree of cleanliness in urban drains and ponds because it is a species which breeds only in small amounts of water. In rural areas water-storage pitchers, discarded coconut shells, or even water-filled climbing steps cut into trees, serve this purpose. In towns the insect turns its attention to similar small accumulations of water in derelict car bodies and in refuse dumps containing tins or car tyres (Fig. 3.6) (Rao *et al.*, 1973). Thus, for *A. aegypti* the efficiency of the town refuse collection service is more significant than the nature of the sewage disposal system. The more refuse which is left uncollected or uncovered, the greater the number of receptacles to catch the rain and provide places for breeding.

## Animals Associated with domestic refuse and waste food

### *Gulls*

Unlike *Aedes aegypti*, most of the other species associated with refuse tips are using them as a source of food. The marked increase in many species of gulls in temperate regions over the last twenty years is largely attributable to their increasing utilisation of food scraps on refuse tips (Murton, 1971).

These town-based gull flocks make nuisances of themselves in a variety of ways. Most important is the hazard they can create at airports. From time to time, civil airliners collide with flocks of birds, often causing engine damage and on a few occasions, serious loss of life (p. 112). It has been estimated that over half these incidents involve gulls. The siting of a refuse tip close to an airport can greatly increase the chance of collisions. A particularly unfavourable arrangement is for a refuse tip and a reservoir to be situated on opposite sides of the airport. The gulls are then likely to commute daily between feeding and resting sites on flight paths which take them across the runways. In a number of countries there are now regulations about the siting of refuse tips designed to minimise this particular hazard.

It has also been suggested that gulls feeding on refuse tips and resting on reservoirs could contaminate drinking water supplies with pathogens such as the salmonellae which cause food poisoning in man. However, salmonellae live for only a short time outside the animal body, so that most of the bacteria would not survive long enough to be transferred to a reservoir on the feet of birds. Gulls also have the habit of picking up large fragments of decomposing animal material and some of these are dropped into reservoirs. This is a more likely source of viable salmonellae, but any modern water-treatment works should be capable of removing these organisms from domestic water supplies.

A quite different facet of the gull problem concerns the conservation of other coastal bird species. Most of the gull species which feed on urban refuse tips breed in coastal areas and there is now good evidence to show that their greatly increased numbers are having harmful effects on other less common species, particularly terns (*Sterna* spp.). Around the coasts of Britain herring gulls (*Larus argentatus*) normally breed on grassy slopes and ledges. As populations have risen they have spilled over into nesting sites such as shingle banks normally used by terns. This change is probably responsible for the elimination of terns from some sites and various attempts have been made to reverse the trend in coastal nature reserves by destroying adult gulls and their eggs (Thomas, 1972). None of these gull problems would arise if urban authorities generally adopted some means of refuse disposal other than tipping. The main alternatives are incineration or some form of reprocessing and composting (Benarde, 1973; Gotaas, 1956). Although either would render the refuse unavailable to gulls, the fertilisers and other materials which can be produced by composting processes are much more useful than the inert clinker resulting from incineration.

### *Other urban bird species*

In temperate regions the other common urban birds are house sparrows (*Passer domesticus*), pigeons (*Columba livia*), and starlings (*Sturnus vulgaris*) (Fig. 3.7). These are scavengers of a different kind; they are basically exploiters of waste food, much of it willingly provided by town dwellers. Pigeons and sparrows are essentially seed-eaters and, in addition to taking urban scraps, make use of pickings from wharves, warehouses and food-stores. For starlings, whose natural diet consists largely of soil invertebrates, towns are less importatnt as a source of food than as sites providing warm winter roosts. This is a major consideration for the starling which is a species poorly adapted to withstand low temperatures.

From time to time a case is made for controlling these birds on the grounds that they represent a health hazard. Certainly a number of fungal pathogens

*Fig. 3.7* Common urban bird species in temperate regions. (*a*) House sparrow (*Passer domesticus*); (*b*) Domestic pigeon (*Columbia livia*); (*c*) Starling (*Sturnus vulgaris*); (*d*) Herring gull (*Larus argentatus*)

can be isolated from bird droppings in towns. One is *Histoplasma capsulatum* which can cause histoplasmosis in man. This is a condition which may involve a local infection of the lungs or a more serious and generalized infection affecting a wide range of body tissues. The other is *Cryptococcus neoformans* which causes infections of the skin, lungs or nervous system, in the latter case producing cryptococcal meningitis. This second species is particularly associated with pigeon droppings and has been identified in almost every city where a search has been made (Randhawa *et al.*, 1965). However, both infections can be contracted in a variety of rural situations and amongst the handful of cases which occur each year only the minority can be linked specifically with towns. Obviously people cleaning out bird roosts are potentially at greater risk and would be wise to wear protective masks. To the public at large the danger seems to be negligible.

There remains one situation where birds are undoubtedly involved in disease transmission. This is in areas of North America where birds such as house sparrows harbour the virus responsible for St. Louis encephalitis (Fig. 3.8). Man can become involved when he is bitten by the mosquito vector which has fed on an infected bird. The first recorded epidemic of St. Louis

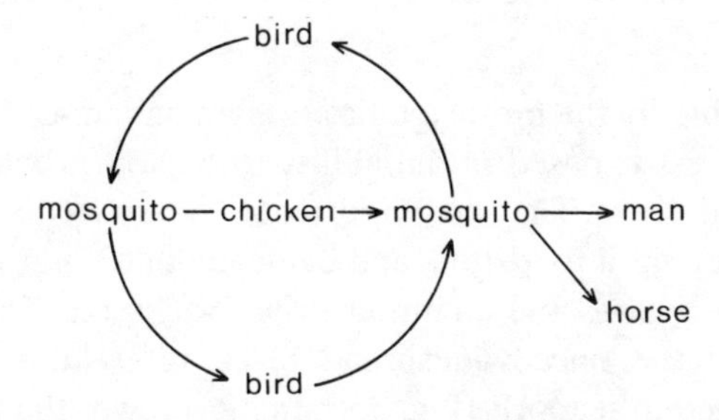


*Fig. 3.8* Transmission chain of St. Louis encephalitis in urban situations. *Culex tarsalis* or *Culex quinquefasciatus* are the mosquitoes commonly involved, the house sparrow (*Passer domesticus*) is an important bird host.

encephalitis occurred in St. Louis County, Missouri, in 1933. Over 1000 people were affected and there were 200 deaths. Since then, in addition to numerous minor outbreaks, there have been serious epidemics in Florida in 1962 and in Houston, Texas, in 1964. In the Houston epidemic 500 adults and children were affected. Mental retardation, paralysis and brain damage with accompanying deformities were some of the conditions that resulted. Nestling sparrows have been clearly implicated as reservoirs for the disease and in these circumstances attempts at bird control can certainly be justified.

A more general objection to starlings and pigeons is that their droppings foul buildings and pavements and may actually destroy the surface ornamentation of soft stonework. Various methods have been used in attempts to discourage the birds from settling on buildings. These have included explosive devices, flashing lights, recorded distress calls, and the application of slippery plastic-gel coatings to ledges. Although these measures can be effective in protecting particular buildings, they normally just divert the problem to another part of town.

Some urban authorities take the view that the cost of protecting and cleaning public buildings more than justifies the institution of measures to reduce numbers. These usually have to be carried out surreptitiously to avoid arousing public opposition. The methods employed include nest-raiding, cage-trapping and the use of narcotising agents such as alpha-chloralose (Thearle, 1968). In North America tests have also been made with baits containing chemicals which inhibit reproduction, but these have the disadvantage of producing side effects in the birds, such as leg paralysis, which would be likely to elicit objections from the general public.

Whatever methods are used, bird control is never likely to be a once-and-for-all operation. Towns are intrinsically so attractive to these species that immigrants from untreated areas quickly move in to replace the individuals which have been removed. This again calls into question the wisdom of embarking on a control programme.

### *Rats*

Rats are undoubtedly the most significant mammalian scavengers in towns. Their great success is based on an ability to exploit urban waste-disposal systems and food stores. They are important as potential disease carriers and also because they spoil foodstuffs and damage buildings.

In temperate regions the common urban rat is the brown rat (*Rattus norvegicus*) and the once widespread black rat (*Rattus rattus*) is now virtually restricted to seaports (Fig. 3.9). In most towns the rat population is

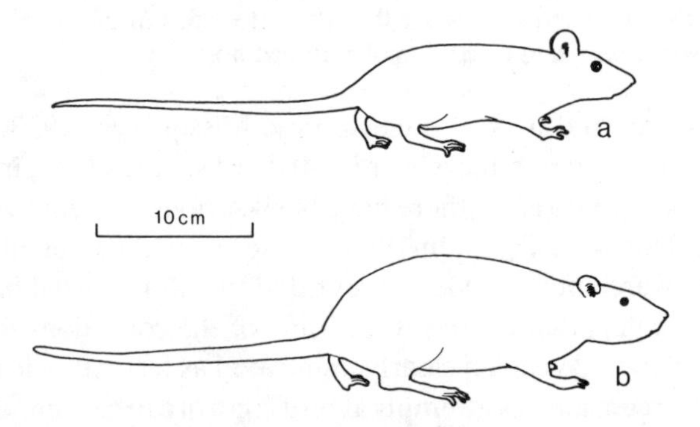


*Fig. 3.9* · The two common urban rats. (*a*) The 'black' rat (*Rattus rattus*) has large ears and eyes and a pointed snout; (*b*) the 'brown' rat (*Rattus norvegicus*) has small ears and eyes and a blunt snout. Colour is sometimes misleading as a distinguishing feature because each species can include black and brown forms

centred on the sewage disposal system and it has been estimated that one km of sewer can support as many as 500 rats. The usual pattern is for the rats to move to the surface at night to search for edible material in food stores and refuse tips and retreat again to the sewers during the day. Even when vigorous control measures are instituted at the surface, the rats can exist perfectly well without ever emerging from the sewers by feeding on material introduced into the system from markets and food premises (Bentley, 1960). Control measures applied simultaneously at the surface and underground are more effective, but even then the benefits are often short-lived because the cleared sections of the sewer system are rapidly recolonised from untreated adjacent areas (Greaves *et al.*, 1968). Because of these habits, there is usually little contact between rats and humans in most modern towns. There is, however, at least one situation involving a potential health hazard. Rats act as carriers of the bacterium *Leptospira icterohaemorrhagiae* which causes Weil's disease or leptospirosis. The organism is excreted in the urine of rats and can infect man by entering the body from contaminated water through cuts and abrasions. In severe cases in which jaundice occurs the mortality rate is high. Rats colonise most open refuse tips and the haphazard dumping of refuse adjacent to a river or lake used for bathing represents a situation of special risk.

*Rats and plague*

Plague is the disease which most people associate with rats and it is interesting to consider why this particular hazard has declined in importance in temperate countries. Over the centuries man has suffered from innumerable local outbreaks of plague and from time to time these have erupted into widespread epidemics. In Europe, major epidemics occurred in Justinian's reign (A.D.542) and in 1348 (The Black Death); in Asia a very serious epidemic occurred at the end of the nineteenth century. It was only during this latter episode that it was finally recognised that plague is essentially a disease of rodents caused by the bacterium *Pasteurella pestis* (*Yersinia pestis*) and transmitted from one rodent to another by the bites of fleas (Hirst, 1953). Man becomes involved when he lives in close proximity to rats with the possibility of infected rat fleas passing the disease to man, particularly after the death of a rat has prompted its fleas to search for new hosts. Man-to-man transmission proceeds either through the activities of man's own flea species or by direct droplet infection (Fig. 3.10). With the benefit of

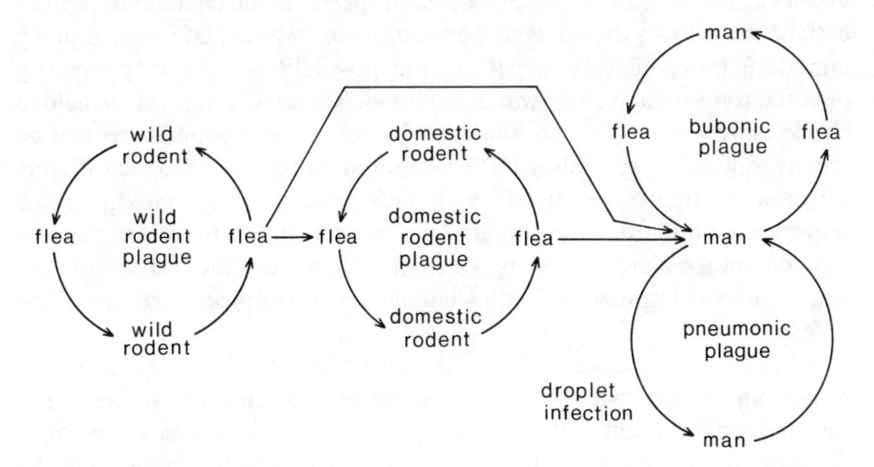


*Fig. 3.10* Transmission chain of plague

hindsight we can appreciate why plague spread so readily in the towns of mediaeval Europe (Shrewsbury, 1971). In the first place the urban rat of this time was the black rat and not the brown rat. The black rat appears to have reached Europe at the time of the Crusades and to have persisted as the dominant species until it was widely displaced by the brown rat in the eighteenth century. The black rat apparently originated as a tree-rat in the tropical or subtropical forests of Asia. In consequence, it is a good climber and has a poor tolerance of outdoor climates in temperature regions.

In mediaeval houses, many of which had thatched roofs with wattle-and-daub or rubble-filled stone walls, the rats would have based themselves in the thatch and used it as a breeding place. One can imagine them making runs in the soft walls and earth floor and using these tunnels at night to exploit the household food store and scavenge in piles of domestic refuse. Because of their sensitivity to cold the rats would rarely have ventured far from the warmth of the building. In this situation the association between rats and people was a close one and when the plague organism reached Europe the results were catastrophic.

It is strange that the significance of the rats was not appreciated during the early epidemics. It was certainly noticed that rats as well as people died, but this was taken as an indication that the 'plague miasma' in the air affected rats as well as man. There was an idea that plague could be carried directly in the fur of animals. During the Great Plague in London this led to the slaughter of domestic dogs and cats which might otherwise have served to check the rat populations.

Many factors have contributed to the virtual elimination of plague as an urban hazard. In Europe the change in rat species in the eighteenth century certainly cut down the contacts between town dwellers and rats. Equally important, however, have been the various possibilities of control opened up once the transmission chain was understood. Rat populations can be held in check using a variety of poisons (p. 57) and the flea populations can be greatly reduced by sprinkling DDT powder in rat runways. Plague sufferers can now be treated effectively with such drugs as streptomycin or the tetracycline antibiotics and contacts can be protected with tetracyclines or sulphonamides. Nevertheless, plague persists in some of the seaports of Asia and became widespread in South Vietnam during the protracted war of the 1960's.

A widespread residual threat remains from the plague infections in wild rodent populations living in many of the major grassland areas of the world. Infected rodents occur in the steppe regions of Asia, the savanna grasslands of South and East Africa, the western prairies of North America and the pampas of South America. It is a matter of argument whether these infections are very ancient or whether they originated during the plague epidemic which spread from Asia at the beginning of the century. Wild rodent plague presents little direct threat to man although trappers occasionally contract the disease from handling dead rodents. More serious difficulties arise when infected wild rodents come into contact with their domestic counterparts and so complete the transmission chain leading to man (Fig. 3.10). In South Africa, for example, plague is well established in the gerbil (*Tatera brantsi*), a burrowing rodent on the sandy veldt. Sometimes the common domestic rodent in South African houses, the multimammate mouse (*Mastomys coucha*), enters gerbil burrows and picks up fleas from animals sick or dead with plague. The fleas are brought back to the house and can give rise to human infection. Elsewhere these unwelcome contacts can occur when species normally living out of doors retreat into houses during flooding or at the end of the crop season when food becomes scarce.

### *Rat control*

The usual approach to rat control involves both preventive measures, such as the modification of buildings to exclude rats and the use of a wide range of poisons (Davis, 1970). Some of these poisons (sodium fluoroacetate, fluoroacetamide, and zinc phosphide) work on a single-dose basis; a single dose being offered to the rats, usually in a cereal bait, after a preliminary conditioning period using unpoisoned baits. Other poisons are designed to be fed more gradually to the rat, a little at a time, until a lethal concentration is reached. The 'anticoagulants' such as hydroxycoumarin (warfarin), coumatetralyl and chlorophacinone, come into this category; they work by interfering with the rat's blood-clotting mechanisms and cause death from bleeding. In recent years rat strains have appeared which are resistant to warfarin (Drummond, 1966). These have been recorded from Scotland, Wales, the west of England, Denmark and Germany and are apparently the result of independent mutations. Special measures have been necessary to control these resistant populations.

There have been a number of novel suggestions for controlling rats by the use of biological agents. These include the use of pathogenic bacteria such as salmonellae and the use of mammalian predators. Both methods have serious drawbacks (Wodzicki, 1973). With bacterial pathogens there is the risk of infecting man or his domestic animals and introduced predators often cause more problems than they solve. For example, the Indian mongoose, (*Herpestes auropunctatus*), introduced into Puerto Rico to control rats has now become a serious reservoir and vector of rabies.

Whatever control measures are used against rats, their association with sewage systems makes it unlikely that they will ever be completely eliminated from towns.

## Pets

In contrast with rats and mosquitoes which are entirely unwelcome as man's associates in towns, there are some animals which are actively encouraged.

Pet dogs and cats are the most notable examples. Both were originally favoured because of their capacity to control rats and mice in food stores and dispose of scraps and refuse. In modern towns these functions have become secondary and most pets now serve principally to provide companionship for their owners.

The fact that pet-keeping serves a useful purpose does not rule out the possibility that it has some unwelcome side-effects. Much attention has been given in recent years to the risk of parasites becoming transferred from pets not only to their owners but also to members of the general public.

*Parasitic worms*

The common tapeworm of dogs and cats, *Dipylidium caninum*, can hardly be regarded as a significant health hazard. It passes its early stages in the flea and is transmitted from one animal to another when a flea is swallowed during fur-grooming. The possibility of a man swallowing a flea is fairly remote but transmission can sometimes occur when a child is licked by a dog which has crushed a flea between its teeth. However, even if such an infection occurs, it rarely produces any symptoms and the parasite can be readily eliminated.

The roundworms, *Toxocara canis* and *Toxocara cati*, are a different proposition. Not only are they readily transmitted to children but there is increasing evidence that they represent a definite health hazard with about 2% of the apparently healthy population showing immunological evidence of past infection. In most cases the parasite produces few symptoms. However, in a few cases parasitic larvae enter vital organs, such as the eye or brain, resulting in loss of vision or epileptic attacks (Woodruff, 1970). There is also some evidence for a link with poliomyelitis. People who have had poliomyelitis also show a higher incidence of past *Toxocara* infections than the population in general and it is suggested that this is the result of migrations of the parasite in the body which break down blood-brain barriers and allow the entry of bacteria and viruses into the nervous system (Woodruff *et al.*, 1966).

On average one in five dogs and cats in Britain is infected with *Toxocara* parasites (Table 3.7) and a similar incidence has been recorded from the United States. The adult parasite lives in the alimentary canal of dogs and cats and in the normal course of events the parasite is spread to other dogs and cats when they swallow eggs derived from faecal masses. Puppies can be infected directly from the mother before birth (Fig. 3.11). Infections of pets by *Toxocara* are sometimes symptomless but often cause intestinal obstructions and the larval migrations of the parasite can cause damage to a variety of organs. When a man picks up the parasite, although it does not grow to

*Table 3.7* Incidence of *Toxocara* infection in dogs and cats in Britain (Data compiled by B. Bisseru)

| | *Dogs* | *No. examined* | *Cats* | *No. examined* |
|---|---|---|---|---|
| Cambridge (1908) | 70·8% | 24 | | |
| London (1922) | 3·0% | 200 | | |
| Aberystwyth (1927) | 16·3% | 59 | 61·8% | 155 |
| Home Counties (1964) | 20.7% | 300 | 22·7% | 176 |
| London (1965) | 6·4% | 250 | 8·0% | 100 |
| London (1966) | 22·9% | 170 | 19·9% | 272 |

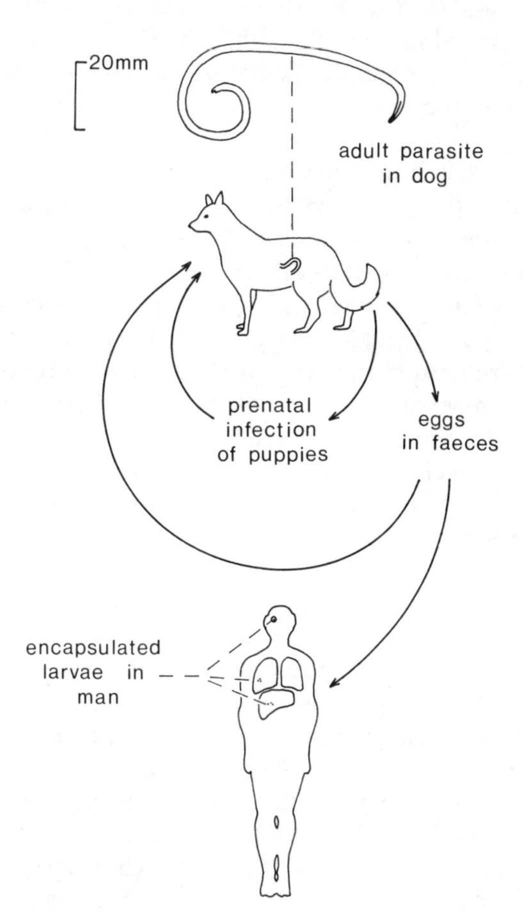


*Fig. 3.11* Life-cycle of *Toxocara canis*

maturity, it can cause a great deal of damage during its larval movements. Particular attention has been given to the loss of vision which can occur when larvae become trapped in the blood vessels of the eye and the resulting lesion causes the retina to degenerate or become detached (Duguid, 1961; Bisseru, 1967).

Toxocaral infections are especially characteristic of young children and there are a number of opportunities for infection from the family dog or cat in the house or garden. However, in at least half the known cases there is no record of a pet in the household. This is a significant finding which suggests that a proportion of infections are picked up in public places. Parks are an obvious possibility and a recent survey of soils from parks in six cities in Britain (London, Birmingham, Norwich, Cardiff, Glasgow and Brighton) showed that there were places in each park where the soil was contaminated with *Toxocara* eggs. In fact, a quarter of the 800 random samples contained eggs (Borg & Woodruff, 1973). Although it is not easy to distinguish between the eggs of cat and dog roundworms, dogs are likely to be more important because they visit parks more frequently and unlike cats do not bury their faeces.

Within the park it is probably not the grassy areas which present the hazard but rather the play areas which have been specially set aside for children and equipped with facilities such as swings, slides, roundabouts and sandpits. In a recent (1976) study of play areas in eight Cardiff City parks, six were found to be fouled by dog faeces containing *Toxocara* eggs. Elsewhere a skin test survey of 65 children using a contaminated play centre showed that seven children had been infected (Borg and Woodruff, 1973).

There are a number of ways to combat this hazard. An obvious one is that children should be taught to wash their hands after playing with pets or in places which are likely to be contaminated. However, this is unlikely to provide an answer with younger children all of whom are prone to suck soiled fingers, whilst some actually eat soil. The problem would be reduced if more pets were regularly dosed with antiworm preparations, particularly those containing piperazine, a drug known to be effective against *Toxocara*. A suitable treatment regime for a dog would be at the ages of three–four weeks, three months, and subsequently every six months. However, unless most owners can be persuaded to treat their pets and a way is found to treat or eliminate strays, then additional methods are needed to protect certain recreation areas. The most immediate need would seem to be to exclude dogs from places such as playgrounds which are extensively used by younger children.

There is at least one other dog parasite which in special circumstances can create a problem in urban areas, the small tapeworm *Echinococcus granulosus*. As an adult it lives in the dog and normally passes its early stages in the form of a large cyst in the sheep. In urban areas where there is a high incidence of straying sheep, the life cycle can become diverted and the cyst develops in man, with serious consequences. A situation of this kind is described later from the industrial valleys of South Wales (p. 189).

### *Rabies*

Rabies is another important disease associated with dogs in towns. It is usually transmitted to man by a bite from an infected dog, the rabies virus entering the wound from the animal's saliva.

Rabies is rightly feared. Not only is there the virtual certainty of death in the absence of immediate remedial treatment, but the disease has very unpleasant symptoms, which include excessive salivation, mental derangement, and an aversion to water (hydrophobia). The usual treatment for a person bitten by a possibly rabid animal has been an unpleasant series of from seven to 21 injections. However, a new vaccine is currently being tested which will involve only a single injection and, if successful, could be used for both preventive and remedial purposes. The magnitude of the rabies problem is indicated by the number of people undergoing preventive treatment each year: in India 100 000 people are treated, in the United States 30 000, and in Europe (excluding Germany) 26 000 (West, 1972).

In urban areas the two control measures which have proved most effective have been the mass vaccination of domestic pets and the destruction of strays. For example in Malaya, where animal health had deteriorated during the second world war, rabies spread widely in the northern parts of the Federation and reached the capital, Kuala Lumpur, in 1952, where 12 000 strays were rounded up and destroyed and a massive programme of compulsory vaccination was instituted involving 18 000 dogs. By the end of the year the incidence of rabies had fallen from 35 cases a month to nil. Similar methods have been applied elsewhere, notably in the United States, and have been generally effective. Difficulties can arise in Buddhist countries where there is a reluctance to eliminate strays and in Moslem countries where dogs are considered unclean and there is a reluctance to handle them for routine vaccination procedures.

Apart from its occurrence in town dogs, it is now known that rabies can occur in a wide range of wild animals including foxes, badgers, jackals, wolves, skunks, mongooses and vampire bats. There are a number of possible links between the urban and rural transmission cycles (Fig. 3.12). For example, an urban dog can be bitten by an infected wild carnivore when taken into the country, or when it encounters a fox scavenging in the town. If rabies becomes established in Britain these urban foxes, which are now such a source of interest to naturalists, will take on a new significance. The

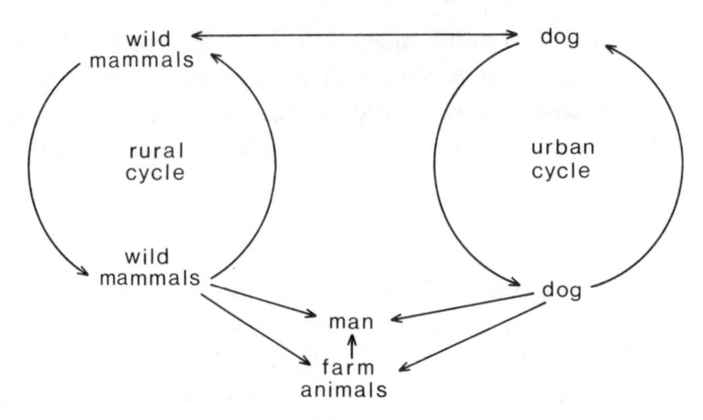


*Fig. 3.12* Transmission chain of rabies

importance of the rural cycle is that it acts as a reservoir from which urban dogs can be reinfected. This means that control measures in towns can never be once and for all operations, but have to be continued indefinitely. The destruction of a large proportion of the potential wildlife carriers in limited areas is technically possible but it is doubtful whether it serves any useful purpose because of reinvasion from surrounding areas.

With all these difficulties associated with rabies, countries such as Britain, Norway and Sweden, which are currently free of the disease, are justified in taking the most stringent measures against its entry. The most important measure is the six-month quarantine period for all dogs and cats and for other wild animals which could act as carriers. This gives time for the disease symptoms to appear whilst the infected animal is in isolation. Since quarantine restrictions were introduced in Britain, 30 cases of rabies have been intercepted in this way. The six-month period is a somewhat arbitrary one and occasionally animals develop rabies after being released from quarantine. In 1969 in Camberley, Surrey, a terrier brought from Germany developed rabies after being released from quarantine. This incident led to a wildlife eradication programme in the area and remedial treatment for 50 people who had been in contact with the dog. A longer quarantine period would intercept the few cases which take a long time to develop but would certainly increase the temptation for owners to smuggle their pets into the country. In 1974, 213 animals were detected being illegally introduced into Britain and this activity is the main threat to our defences against rabies.

Rabies has spread westwards in Europe during the last 30 years. It crossed the Oder River from East to West Germany in 1947 and crossed the Rhine into France in 1968. In France it has spread in a southwesterly direction at the rate of about 50 km a year and is likely eventually to reach the French Channel ports. This will increase Britain's need for effective antirabies defences and to this end recent legislation has increased the penalties for introducing animals into the country illegally. Britain is fortunate in having the Channel as a defence line. A rabies-free country which only has a land boundary to separate it from an infected area is in a more difficult position because of the problem of preventing the free movement of wild carriers such as foxes.

In spite of the problems, pets serve an important function in towns. The fact that they can transmit diseases to man in no way establishes a general case for not keeping pets. Indeed, if we looked after our pets more carefully we would go a long way to safeguarding our own health. If all dogs were protected against worm infections and (in the appropriate countries) immunised against rabies, and if dogs were never abandoned to become strays, the hazards to human health would virtually disappear. Unfortunately, a universally high standard of care is no more likely amongst owners of pets than amongst parents of children, and in these circumstances urban authorities need to devise general measures to eliminate strays and protect recreation areas used by young children.

# 4 Industrial development

Industrial development can interfere with other interests in a variety of ways. Objection is raised to many industrial projects on the grounds that they are unaesthetic, and in recent years considerable progress has been made in the development of landscaping techniques to camouflage industrial sites. One of the ecologist's particular roles in this connection has been to identify the factors which impede the growth of plants on various industrial wastes and to suggest ways to overcome these limitations. In recent years a number of comprehensive compilations of information in this subject have been published (Hutnik & Davis, 1973; Goodman & Chadwick, 1975; Goodman & Bray 1975) and only a few examples need to be mentioned here.

With some industrial wastes such as the spoil from kaolin extraction nutrient deficiency is the main impediment to plant growth. In these circumstances the combined application of fertilisers and appropriate seed mixtures is all that is necessary to establish a cover of vegetation. At the other end of the scale are derelict mine and smelter sites, where not only are the wastes infertile, but they are also likely to contain toxic substances. Whilst it would be theoretically possible to cover such material with a deep layer of soil, topsoil is, in general, too valuable to use for this purpose. Nor is there any certainty that the toxic elements would remain below the surface. The alternative strategy for revegetating these sites is to use plants which have developed a natural tolerance to heavy metals. A large number of such species are known from all over the world. Many are so specific in their association with naturally occurring metal-contaminated soils that they have been used by prospectors in their search for ore bodies (Cannon, 1971). For example, in Australia the grass *Eriachne mucronata* is associated with lead and in Africa the basil, *Becium homblei*, with copper deposits. Some species have a normal form, characteristic of uncontaminated soils, and a different strain or variety with metal-tolerant properties. In Western Europe this is the case with the mountain pansy *Viola lutea* which has a variety *calamineria*, tolerant of high zinc concentrations. Of particular importance for practical purposes is the fact that many common grasses are now known to have metal-tolerant strains. By accumulating seed stocks from such strains and sowing them, in association with fertilisers, on mine and smelter wastes it has proved possible to establish a vegetation cover on sites where natural colonisation would have been minimal (Gadgil, 1969; Bradshaw, 1970; Antonovics, *et al.* 1971). It should be emphasised however, that this is essentially a landscaping operation. The grass cover so formed, although it may look quite normal, is unlikely to be suitable for grazing by animals because of the accumulation of heavy metals in the plant tissues.

Apart from aesthetic considerations, industrial development frequently involves the physical destruction of sites valued for other purposes. In the case of some conservation sites, for example where limestone quarrying destroys a cave system, or lime extraction a coral reef, the process is irreversible. In these cases the ecologist's role must be to provide an objective grading of potentially threatened sites, which can be assessed against industrial priorities. In a few instances the destruction of one type of habitat eventually results in the creation of another. The best example of this is the case of gravel extraction where the water-filled pit remaining at the end of the operation can represent an important extra habitat for animals, particularly wildfowl, and for aquatic plants (Catchpole & Tydeman, 1975; Davis, 1976).

Of all the effects of industry, pollution is the most pervasive. Many of the substances we have come to regard as industrial pollutants were in fact being liberated into the environment from natural sources long before the industrial revolution. In Iceland, for example, the fumes from the volcano Hekla were recorded in the eleventh century as causing the kind of damage to farm stock which we have since come to associate with fluoride emissions from aluminium smelters and fertiliser factories. Similarly many coastal waters were being affected by natural seepages of oil through permeable strata long before the occurrence of spillage from drilling mishaps and tanker accidents. For such pollutants, the main effect of industry has been to multiply the sources of emission. In a quite different category are substances such as polychlorinated biphenyls (PCBs) which did not exist before they were synthesised by man, but are now widely dispersed in the environment. (Edwards, 1971).

Pollutants from industry can interfere with many interests and enterprises. In the field of biological conservation for example, oil-spillage represents a significant threat to groups of sea birds such as diving ducks, auks and penguins (Croxall, 1975). The Torrey Canyon disaster alone is estimated to have killed something like 40 000 auks, and with some localised species such as the jackass penguin (*Spheniscus demersus*) living in scattered colonies around the South African coast, oil pollution appears to have been a major factor causing a progressive population decline (Frost *et al.*, 1976). There are also examples of pollution threats to other habitats. In recent years there have been fears that the discharge of pulp-mill effluents into Lake Baikal in southern Siberia could affect the lake's unique plant and animal populations (Pryde, 1972). On the other side of the world, in America's Glacier National Park, fluoride emissions from an aluminium smelter have been shown to be producing extensive biological damage.

Of more direct relevance to man is the possibility of industrial pollutants causing damage to his own health, to agricultural and forest crops, and to fisheries.

## Pollution effects on agriculture, forestry and fisheries

Where industrial emissions are causing damage to crops or farm stock, the severity of the symptoms is often related to distance from the emission source. This is well-illustrated by the effects of fluoride on farm stock. In a study of the area around Stoke-on-Trent, the centre of Britain's ceramic industry, it was possible to distinguish two levels of damage in cattle. (Burns & Allcroft, 1964). In an inner zone, close to the sources of emission the animals showed severe symptoms of fluorosis including skeletal malformations and lameness (Fig. 4.1). Further out, symptoms were restricted to tooth staining and minor tooth damage of little functional significance. Eventually the point was reached between three and six kilometres from the emission sources (dependent upon the prevailing wind direction) where the herds were free of fluoride damage of any kind (Fig. 4.1). Some of the typical symptoms of animal fluorosis are shown in Figure 4.2*a*,*b*.

A similar type of concentric pattern is evident with fluoride damage to forest trees. Gilbert (1975) has described how, in the vicinity of Norwegian aluminium smelters, there is an inner zone where all the coniferous trees have been killed, and an outer zone where a proportion of the trees have survived. As might be expected, these damage patterns were related to the size of the smelter, local topography and prevailing wind directions. In a valley site, for example, the pattern was elongated along the axis of the valley in both directions. Foliage damaged by fluoride shows characteristic browning towards the tip of the leaf with a conspicuous reddish-brown line dividing the damaged from the healthy area (Fig. 4.2*c*,*d*).

In addition to fluorides, a number of other industrial emissions have been shown to be harmful to crops and stock. Horses are especially vulnerable to herbage contamination from lead smelters (Hammond & Aronson, 1964, Schmitt *et al.*, 1971) and there are instances of cattle, sheep and horses being poisoned by eating grass contaminated by arsenic trioxide from copper smelters (Harkins & Swain, 1908). Copper smelters are also a major source

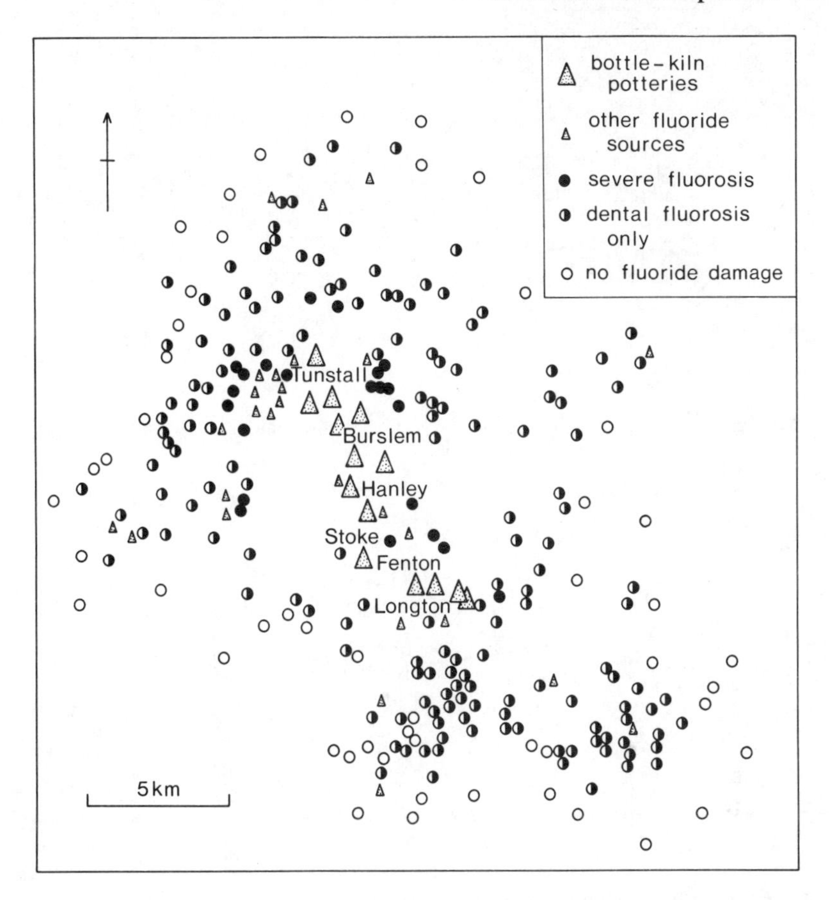


*Fig. 4.1* The incidence of cattle fluorosis in the Stoke-on-Trent area in relation to sources of fluoride emission 1954–57 (after Burns & Allcroft, 1964)

of damage to commercial forest areas, in this case the harmful components of the emission being copper and sulphur dioxide. The effects of sulphur dioxide pollution from smelters can be very far reaching. Around the large smelting complex at Sudbury in Canada, pollution levels likely to cause damage to forest trees have been recorded over an area of 5470 $km^2$ and at distances of up to 45 km from the point of emission. (Hutchinson & Whitby, 1974).

Feasibility studies for new industries need to take account of the fact that the discharge of effluents to the atmosphere can render the surrounding area unsuitable for agricultural and forestry use. To predict the extent of possible damage, information is needed both on the sensitivity of plant crops or animals to pollutants, and also on the likely dispersal pattern of effluents from the point of emission. These patterns are determined not only by total

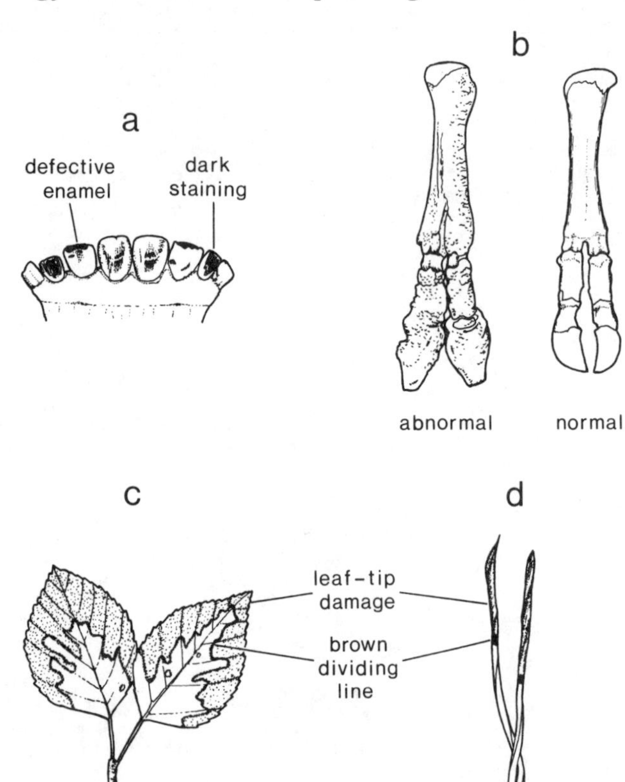


*Fig. 4.2* Symptoms of fluoride damage: (*a*) Incisor staining and malformation and (*b*) Excessive ossification of lower limbs in cattle. Foliage damage in (*c*) Birch (*Betula pubescens*) and (*d*) Pine (*Pinus sylvestris*) (*c* and *d* after Gilbert 1975)

emission levels but also by such factors as chimney height, prevailing wind direction and topography (Fig. 4.3).

The prediction of biological effects may be made more complicated by interaction between one pollutant and another. Sometimes the interaction increases the amount of damage caused, sometimes it decreases it. For example, the effects of sulphur dioxide can be increased by the presence of dust because dust particles wedge open the leaf stomata and prevent them from closing normally at night. This can so increase the total entry of sulphur dioxide into the leaf that photosynthetic pigments are extensively damaged. This in turn leads to premature leaf fall and a reduced growing period (Ricks & Williams, 1974, 1975). The damaging effects of sulphur dioxide may also be augmented by pollutants such as ozone and nitrogen

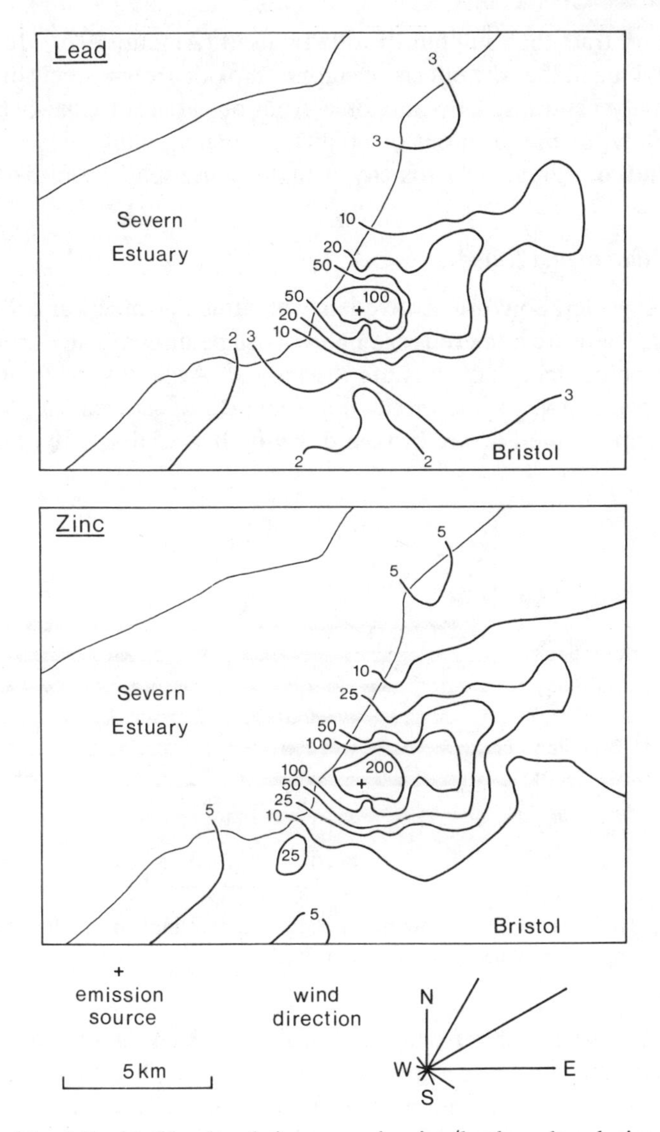


*Fig. 4.3* The fall-out of lead and zinc around a zinc/lead smelter during a 21-day period, in relation to wind direction. Suspended moss bags were used for sampling and the concentration contours represent $\mu$g/g moss/day (after Little & Martin, 1974)

oxides associated with photochemical smog (Menser & Heggestad 1966; Tingey *et al.*, 1971). In contrast the damaging effects of sulphur dioxide on animals can actually be reduced by the presence of ammonia. The ammonium sulphate which is formed is considerably less harmful to the respiratory tract than sulphur dioxide by itself (Amdur, 1971). In practice this could mean that the effects of sulphur dioxide from a steelworks could be partially neutralised by ammonia from an adjacent coke-oven plant. Fortunately for the purposes of prediction, many pollutants act independently and contribute to the toxicity of mixtures in a simple additive fashion.

*River pollution and fisheries*

Similar considerations are involved in evaluating the effects of pollution on fisheries. There are numerous examples of fish being eliminated from a river below a point where effluents are discharged. According to their various degrees of sensitivity some species may reappear downstream as the pollutants become diluted or are broken down by bacterial activity. Figure 4.4

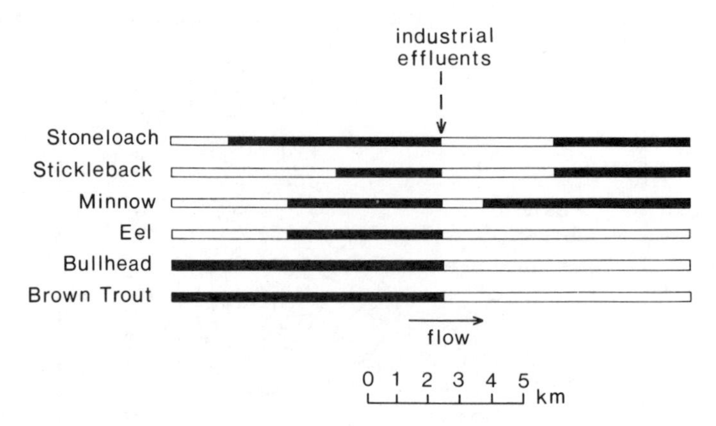


*Fig. 4.4* The effect on fish populations of the entry of industrial effluents into the River Cynon, South Wales (after Learner *et al.*, 1971)

shows a typical example from South Wales in which the discharge from a smokeless fuel plant, consisting mainly of cyanides, phenols and ammonia, eliminated brown trout (*Salmo trutta*), eels (*Anguilla anguilla*) and bullheads (*Cottus gobio*) from the river and produced a localised disappearance of minnows (*Phoxinus phoxinus*), sticklebacks (*Gasterosteus aculeatus*) and stoneloach (*Noemacheilus barbatulus*). In this situation only the trout are of direct interest to anglers.

As with air pollutants, some pollutants in aquatic systems act together aggravating one another's effects. Examples include the combination of cyanides with ammonia, or cyanides with nickel or silver. However, many combinations act in a relatively straightforward additive fashion. For example laboratory experiments have shown that either 4 mg/l of phenol or

2 mg/l of zinc added to a tank of rainbow trout (*Salmo gairdneri*) will kill 50 per cent of them in 48 h. If the two substances are combined, the same mortality is produced by 2 mg/l of phenol plus 1 mg/l of zinc. (Herbert & Vandyke, 1964). In other words, a combination of two half-toxic doses of each pollutant has the same effect as a whole toxic dose of either separately. Moreover experiments with combinations of (i) phenol, zinc and copper, and (ii) nickel, zinc and copper suggest that the same approach can be applied to mixtures of three or more pollutants, at least where no single constituent is making the overriding contribution to total toxicity (Brown & Dalton, 1970). There is, of course, a basic difference between the situation in a laboratory tank with fixed concentrations of pollutants, and the fluctuating conditions encountered by fish in a river. Nonetheless recent observations on the River Trent system in Britain suggest that the results of laboratory toxicity tests can be used to predict whether a river section with particular chemical characteristics is likely to support fish (Alabaster *et al.*, 1972). Ultimately this approach should make it possible to predict whether a specified pollutant input from a new industry is likely to eliminate fish, or conversely what degree of improvement in effluent quality is necessary for fish to return to a fishless zone. A comprehensive account of the causes and effects of river pollution is given by Klein (1962).

## Possible beneficial effects of industrial effluents

With so much attention being given to the damaging effects of industrial effluents it is easy to lose sight of the fact that effluents can sometimes have beneficial side effects. Low levels of atmospheric contamination by sulphur dioxide, nitrogen oxides and ammonia may actually have an enriching effect on some forest and farm soils. In aquatic situations hot water is the prime example of a waste material which with a little ingenuity can be made to serve a useful purpose.

Many industrial enterprises, but particularly electricity-generating stations take water for cooling purposes from rivers, lakes or the sea and then return it at a somewhat higher temperature. The temperature rise is usually of the order of 10°C. Much attention has been given to the possibility of these heated discharges causing damage to fisheries (Alabaster, 1963; Clark, 1969). River fish sometimes die as a result of being trapped in water-outlet areas and occasionally, off the Atlantic coast of North America, appreciable numbers of menhaden, (*Brevoortia tyrannus*), a fish of commercial importance, have perished as a result of swimming into undispersed layers of hot water offshore (Young & Gibson, 1973). However in view of the number of operational power stations the incidence of these problems is remarkably low.

Every aquatic species has an upper temperature limit (the upper lethal limit) beyond which it cannot survive (Fig. 4.5). As might be expected this limit is higher for tropical or sub-tropical species than for temperate ones. In the few cases where fish have been killed by heated effluents it is because the temperature of the effluent has exceeded the upper lethal limit of the species involved. Within the range that a species can tolerate it is also possible to define a more limited zone which is optimum for growth (Fig. 4.5). It is the

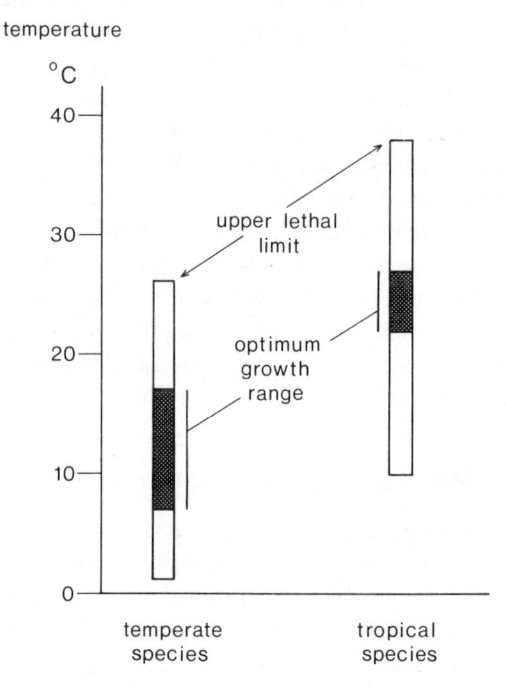


*Fig. 4.5* Typical temperature tolerance and optimum growth ranges in temperate and tropical species

position of this optimum growth range which holds the key to the utilisation of heated effluents. Many temperate species living in normal outdoor environments have their growth checked when temperatures fall in winter. By keeping such species in tanks or lagoons warmed by heated effluents it is possible to eliminate this seasonal check and increase overall growth. It may also be possible to accelerate growth by keeping the temperature near the upper end of the optimum growth range. If the amount of waste heat available is large it may be preferable to make use of tropical species such as the fish *Tilapia*.

There have been a number of pilot schemes for rearing commercially important species using this approach. In American experiments it has been shown that the rearing time for the American lobster (*Homarus americanus*) can be reduced from six to two years by keeping the animals continuously at 22°C. At this temperature eggs hatch more quickly, moulting occurs more frequently and the overall growth rate is accelerated. Various shrimp and lobster rearing projects have been started in Japan, North America and the United Kingdom. The use of heated ponds for fish-rearing is a well-established practice in Poland and the U.S.S.R. The carp (*Cyprinus carpio*) is the favoured species and it has been established that the pond temperature needs to be kept within the range 25–27°C for maximum growth. In Britain, the most significant fish-rearing projects have involved plaice (*Pleuronectes platessa*) and sole (*Solea solea*), two marine flat fish for which there is a ready commercial outlet. Optimum growth temperatures have been identified as 15–16°C for plaice and 18–20°C for sole. By manipulating the heated effluents at two power stations to produce these conditions in lagoons it has proved possible to grow fish to marketable size in 18 months, rather than the 30 months which is the time taken in natural situations (Nash, 1968).

Providing food for the culture animals often accounts for about half the cost of these projects and many will become commercially feasible only if cheap food sources are found. Food processing wastes and sewage sludge are obvious starting points for such materials and the whole question represents an important challenge in the field of effluent and waste management.

## The effects of industrial pollutants on human health

Harmful materials of industrial origin can enter the human body by a variety of routes; in drinking water, in food, in inhaled air and through the skin. Some hazards are virtually restricted to industrial working environments; the various dust diseases (asbestosis, pneumoconiosis etc) are obvious examples. Of more direct relevance to land-use planning are the situations where industrial emissions represent a localised hazard to the general public.

There are a number of instances of localised health effects which are directly attributable to the inhalation of pollutants from the atmosphere. Long term exposure to airborne beryllium can cause respiratory distress and the formation of granular growths in the lungs. In a number of cases in the United States these symptoms have been found in people living as far as 10 km downwind from a beryllium-producing plant (Sussman *et al.*, 1959). Health hazards can also be created by the fall-out of particulate matter from manganese factories. In the early 1930s, a high incidence of pneumonia in the Norwegian town of Sauda was eventually traced to atmospheric pollution from this source. More recently in 1962 in Aosta, Italy, the occurrence of ear, nose and throat problems amongst children living within a 500 metre radius of a manganese factory was attributed to airborne contaminants (Waldbott, 1973).

The Minamata episode in Japan provides one of the most clear-cut examples of a poison of industrial origin being ingested in food. Between 1953 and 1960 a strange illness led to the deaths of 43 people and the serious disablement of 75 others in the vicinity of the coastal town of Minamata (Fig. 4.6). The symptoms which included serious disorders of vision, loss of muscular co-ordination and other neurological disturbances were mainly apparent amongst fishermen and their families. Eventually the cause of the problem was traced to the discharge of mercury from a plastics factory. Mercury was being used as a catalyst in that part of the process concerned with aldehyde production, and mercury rich wastes were being discharged, first into the Minamata Bay and subsequently, in an attempt to increase the dilution, into the Minamata River. Fish and shellfish from these waters were found to contain high levels of mercury in their tissues and were obviously the source of the poisoning. Another epidemic of the same disease occurred

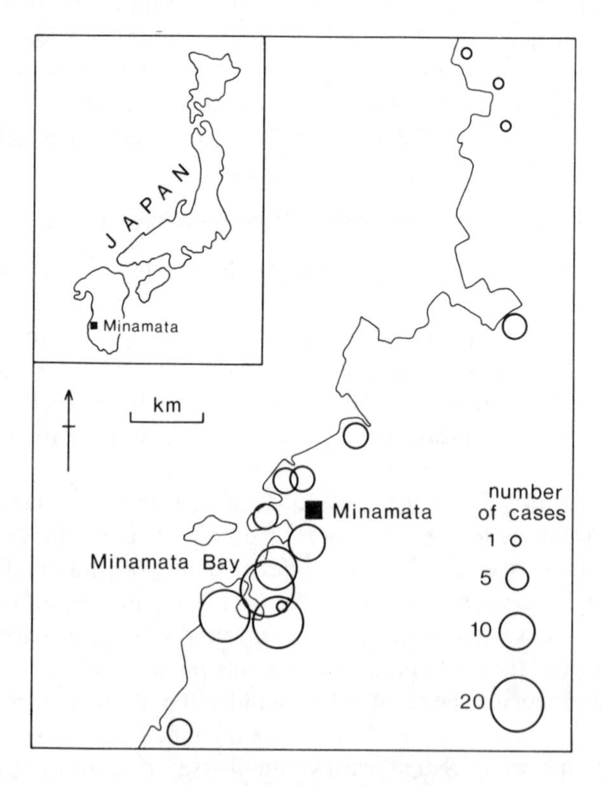


*Fig. 4.6* The incidence of mercury poisoning in villages near Minamata (based on data in Tucker, 1972)

in 1965 in Niigata on the Japanese island of Hon Shu. This again involved the contamination of fish by mercury from a plastics factory.

Before these episodes, no great risk was felt to be involved in the discharge of inorganic mercury to watercourses. At the Minamata factory, mercury was apparently not even included in the factory's routine sampling of effluents. It only became apparent from subsequent work in Scandinavia (Jensen & Jernelöv, 1969) that aquatic micro-organisms are capable of converting inorganic mercury into a much more dangerous organic (methyl) form. By accumulating in the flesh of fish this can represent a direct health hazard to man.

Japan provides another notable example of food contamination from a local industrial source (Emmerson, 1970). For many years people living in the Fuchi-machi farming community at the edge of Toyama City had been prone to a disorder involving severe bone pains and a tendency to multiple bone fractures. The pains were so severe that the disease became known locally as 'itai-itai byo' which literally translated means 'ouch-ouch' disease. When autopsies showed unusually high cadmium concentrations in body tissues, a search for its source revealed high levels of the metal in local rice and soybean crops. The cadmium was shown to originate from a nearby lead – zinc mining and smelting complex. This was contaminating, with particulate cadmium, the stream used to irrigate the rice-growing areas. Subsequently, another focus of 'itai-itai' disease has come to light in Japan near Bandai, again involving the contamination of rice with cadmium from a zinc smelter. This regular association of cadmium emissions with zinc smelters results from the natural occurrence of small amounts of cadmium in zinc ores.

Some health problems are the result of toxic materials entering the body by a variety of routes, not only by inhalation from the air but also by ingestion in food and drinking water. Many examples of localised fluoride poisoning are of this type. One of the first scientific studies was made in the ironstone mining region of South Lincolnshire (Murray & Wilson, 1946). Here it was the practice to prepare the ore for smelting, by roasting it with coal in heaps alongside the mine sites. The fluoride-rich fumes from this operation were shown, in the case of one farm, not only to be causing lameness and poor growth in sheep and cattle but also to be responsible for ill-health in the farmer and his family. The symptoms, now known to be typical of so-called 'neighbourhood' fluorosis in humans, included gastric and respiratory disorders and bone pains. More recently Waldbott and Cecilioni (1969) have described a similar situation on a fruit farm in Ontario following the establishment of a superphosphate fertiliser factory about half a kilometre away. Fluoride emissions caused progressive deterioration in the strawberry and rhubarb plants, currant bushes and fruit trees on the 20 hectare farm. In the garden plot, burns appeared on the tips and margins of begonia and geranium leaves and 83 of the farm's 86 colonies of bees were wiped out. Over the same period the farmer's wife and son developed the characteristic symptoms of human fluorosis. In such circumstances plant and animal damage can provide an important early warning of threats to human health.

The prediction of human health hazards can be approached in much the same way as predictions about animal and crop damage, and a number of useful summaries are now available both on the nature of the waste emissions associated with different industries (Bond & Straub, 1973 – 74), and the implications for health (Waldbott, 1973). Whilst, given the necessary information, it may be feasible in land-use planning to take account of routine effluent emissions, it is much more difficult to make allowance for the possibility of accidental discharges.

A small proportion of the mishaps and malfunctions at industrial installations result in the release of dangerous materials into watercourses and into the atmosphere. The accidental release of hydrogen sulphide at Poza Poca in Mexico's biggest natural gas field affords a typical example. A new plant designed to remove hydrogen sulphide from natural gas and convert it into commercially useful sulphur had been opened there. As a result of a fault in a pipeline valve, hydrogen sulphide escaped into the air of the town for about 25 min. A temperature inversion caused the escaped gas to accumulate in a sufficient concentration to cause the death of 22 people and illness in another 320. In high concentrations, (in excess of 1000 $mg/m^3$), hydrogen sulphide acts by paralysing the nerve centre controlling respiration and as a result is rapidly fatal. In this particular incident, the gas also killed about half the domestic animals in the area, including chickens, cattle, pigs, geese, dogs and cats.

A more recent incident has been the release of toxic material from a chemical plant at Séveso in northern Italy (Hay, 1976a,b). The plant was designed to produce trichlorophenol, a substance used in the manufacture of the herbicide 2,4,5-T and the anti-bacterial agent hexachlorophene. A highly toxic compound dioxin is a by-product of trichlorophenol production. On July 10th 1976, a runaway reaction in one of the reaction vessels at the factory blew a safety valve and released a hot vapour into the atmosphere. This vapour deposited dioxin (and trichlorophenol) to the south of the factory over an area of 30 – 40 hectares populated by some 2000 people. 500 people have since been treated for dioxin poisoning the symptoms of which include kidney and liver malfunction, foetal damage, reduced white cell counts and acne. The toxin has penetrated sufficiently deeply into the soil (at least 25 cm) to raise doubts about the prospect of decontaminating the site and making it suitable for habitation.

Such incidents suggest that accident-prone industries, quite apart from being subjected to rigorous supervision should also be automatically separated by a buffer zone from centres of population.

# 5 Man-made lakes

Artificial lakes are constructed principally to generate power and to store the water needed to meet agricultural, urban and industrial requirements. Unfortunately, the benefits derived from impoundment schemes are frequently marred by unwelcome side-effects. Some of these secondary consequences are obvious. Most major projects have involved the displacement of human settlements and many have adversely affected agricultural areas and sites of biological and archaeological interest. Salvage operations have included the planned evacuation and resettlement of human communities, the evacuation of game animals at Lake Kariba and the removal of the Temple of Abu Simbel at the site of Lake Nasser. In temperate regions it has been realised for many years that dams need to be specially modified to prevent interference with salmon migration.

As more dams are built, the list of possible side-effects increases. For example, it is now known that large impoundments can increase earthquake activity and tremors have occurred at a number of sites including Lake Kariba. These are caused by the great weight of water in the lakes disturbing the established tectonic equilibrium. On the biological side, it is becoming apparent that most major impoundments in the tropics are likely to have significant effects on the health of local communities, typically increasing the incidence of those diseases whose vectors or intermediate hosts thrive in still-water conditions. Downstream impacts are also receiving more attention; impoundments smooth out or 'regulate' the pattern of floods in the river below and can affect farming, fishery and conservation interests for hundreds of kilometres downstream. Not all the side-effects of lake construction are necessarily detrimental, a properly managed fishery on a lake can represent an asset to a protein-hungry population and, in more highly developed societies, lakes can help to meet the growing demand for recreational facilities.

There is an increasing awareness of the importance of these secondary effects and a number of symposia have been organised recently to promote this wider approach to the planning and management of artificial lakes (Lowe-McConnell, 1966; Obeng, 1969; Stanley & Alpers, 1975).

The specifically biological contribution falls into two parts. One concerns the effects of lake construction on other interests in an area, and some examples of this kind of interaction are listed in Table 5.1. The other involves the management of the lake itself. Sometimes biological events at a newly-constructed lake threaten to interfere with its primary functions of power generation or water supply. Massive growths of weeds can block the turbine intakes on hydroelectric dams and dense algal growths sometimes make the water in reservoirs unusable for domestic and industrial purposes. The management problems will be examined first.

*Table 5.1* Possible biological impacts of man-made lakes

| *Interest affected* | *Impacts at site of impoundment* | *Impacts downstream* |
|---|---|---|
| Health | – increase in schistosomiasis associated with spread of snail populations<br>– increase in mosquito-borne diseases | |
| Fisheries | – interference with passage of migratory fish<br>– loss of river fisheries<br>– creation of conditions suitable for lake fisheries | – reduction of stimuli for upstream migration of salmonid fish<br>– interference with downstream flood fisheries |
| Agriculture | – inundation of agricultural land<br>– creation of opportunities to cultivate draw-down area of reservoir | – interference with agricultural practices based on seasonal flooding |
| Biological conservation | – inundation of terrestrial sites of conservation interest<br>– interference with estuarine habitats by modification of tidal range or inundation with freshwater<br>– creation of new habitats for wildfowl | – alteration of the flood regime of wetland sites |

### Problems caused by water-weeds and algae

Both microscopic algae and larger macrophytic weeds may form troublesome plant growths in lakes. One of the best-known examples was the invasion of Lake Kariba reservoir by the water-fern (*Salvinia molesta*) (Fig. 5.1*a*). The weed appeared first in the centre of the lake about six months after the dam was closed and subsequently spread around the shore line. After two years (1960) it had covered an area of 420 km$^2$, 10 per cent of the

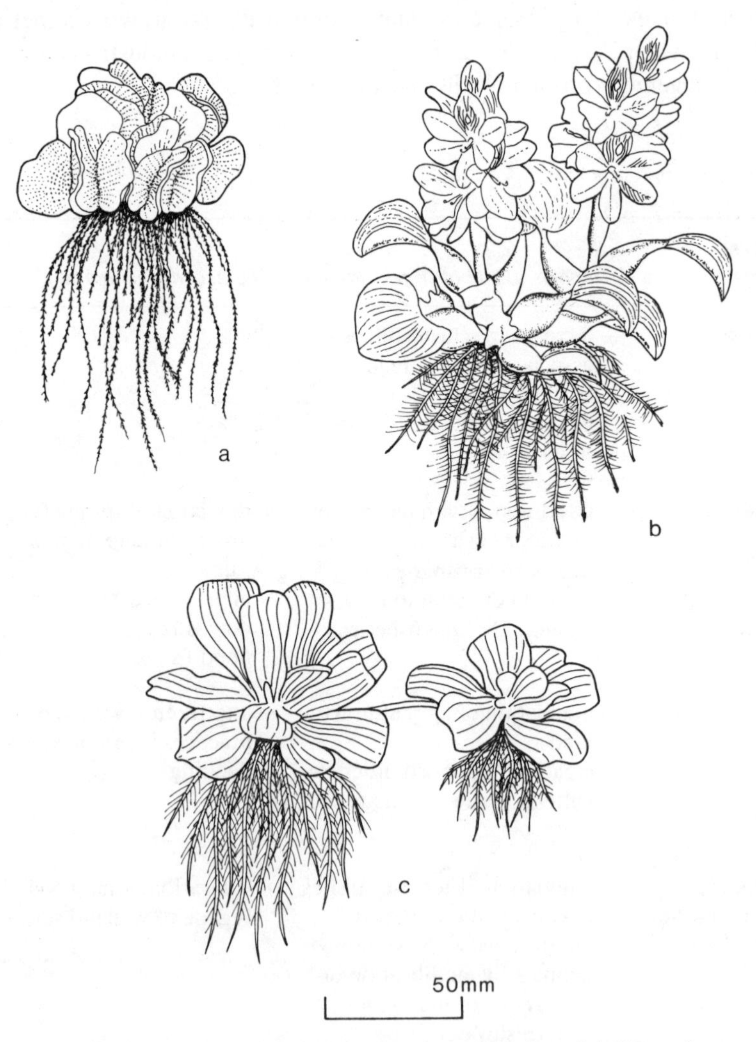


*Fig. 5.1* Some important water-weeds of man-made lakes: (*a*) Water-fern (*Salvinia molesta*); (*b*) Water-hyacinth (*Eichhornia crassipes*); (*c*) Water-lettuce (*Pistia stratiotes*)

lake surface, and after a further two years the area infested had reached 1000 $km^2$ (Mitchell, 1974).

The weed growth caused several problems on the lake. Fishermen working from dugout canoes had difficulty in paddling their boats through the weed masses and found that the growth interfered with the setting and lifting of their nets. Where the mat had been colonised and consolidated by other plants, even larger boats, up to 15 m long with powerful marine diesel engines, were unable to make progress. At the top of the lake, compacted weed masses were periodically blown into Devil's Gorge causing the hydrofoil passenger service to be suspended. Fortunately the plants did not interfere with power generation because the turbine intakes are 20 m below the surface and the prevailing winds tend to blow the weed masses to the head of the lake rather than towards the dam. By the time the lake had reached its final water level in July 1963, it was evident that the outburst of *Salvinia* had passed its peak and was starting to decline. Some submerged weeds such as *Ceratophyllum demersum* and *Potamogeton pusillus* subsequently increased in abundance but apart from their significance in supporting disease-carrying snails these plants caused no major disruption of activities on the lake.

The circumstances which favoured the water-fern outburst at Kariba were not clear at the time and are still a matter for speculation. In more recent hydroelectric schemes, such as the ones at Volta and Kainji (Fig. 5.2), the filling of the lakes was watched with some concern to see whether the pattern would repeat itself. At neither site however has any significant weed problem developed. On Lake Volta, rafts of water-lettuce (*Pistia stratiotes*) (Fig. 5.1*c*) and growths of the submerged weed *Ceratophyllum demersum* appeared soon after the dam was closed in 1964, but these have remained limited in extent and confined to sheltered bays and mouths of inflow rivers. At Lake Kainji the development of weeds has been similarly restricted.

Various suggestions have been made as to why Lake Kariba should differ from the other lakes. There is little reason to believe that it is intrinsically more productive. All three lakes undergo the seasonal mixing which brings a new supply of nutrients to the surface waters and is a prerequisite for high productivity in a tropical lake (Beadle, 1974). Kariba was different in one respect however, in that it was the only site where there was a programme of tree-felling and burning on the area to be flooded. This was undertaken to prevent the submerged trees from interfering with fishing nets, but was not repeated to any extent at Volta and Kainji for financial reasons. It is possible that nutrient-rich ashes from the burning operation helped to promote the initial weed outburst at Kariba.

Some people have attached more importance to the fact that *Salvinia* is an alien species in Africa and might therefore be expected to flourish in the

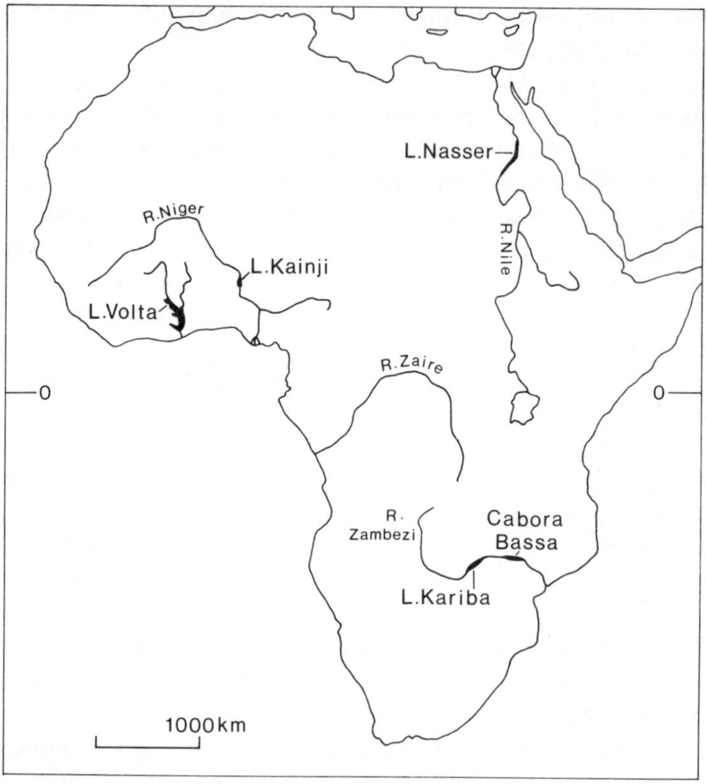

*Fig. 5.2* The principal man-made lakes in Africa

absence of its usual enemies and competitors. The fact that *Salvinia* was already present in the River Zambezi when the Kariba Dam was built, but had not reached the Niger or Volta catchments, would on this hypothesis be sufficient to account for their different histories. If the exotic nature of the plant is indeed the key factor then further difficulties can be anticipated at some of the other new dam sites in Africa. The Cabora Bassa dam on the Zambezi and the Inga dam on the River Zaire come within the water-fern's present range (Little, 1965) and will provide interesting, although possibly inconvenient, large scale experiments.

The water-fern is not the only floating plant to produce troublesome growths in impoundments. The water-hyacinth (*Eichhornia crassipes*) (Fig. 5.1*b*) has created problems in many tropical and subtropical areas, including the Brokopondo Reservoir (Surinam) in its native South America.

### *Problems associated with algae*

In temperate regions the most widespread aquatic plant problems are caused not by the larger weeds, but by microscopic algae. The blue-green algae, *Anabaena*, *Anacystis* (*Microcystis*) and *Aphanizomenon* are some of the most important (Fig. 5.3). They can accumulate so rapidly in the surface waters of a reservoir that a scum or 'bloom' resembling thick pea soup is formed. Besides the blue-green algae, blooms can also be produced by the diatoms *Fragilaria* and *Melosira* and the flagellates *Dinobryon* and *Synura*.

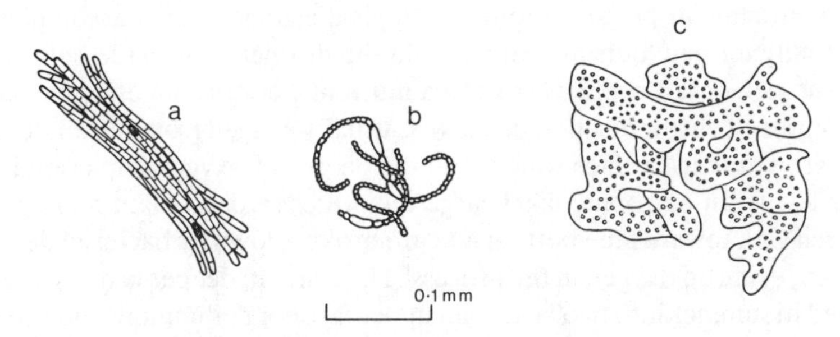


*Fig. 5.3* Blue-green algae commonly involved in bloom formation: (*a*) *Aphanizomenon*; (*b*) *Anabaena*; (*c*) *Anacystis*

These various algal growths can interfere with water supply reservoirs by imparting unpleasant tastes and odours to the water. Putrefying odours are usually associated with large concentrations of blue-green algae and fishy odours with many flagellates and diatoms. Even a small growth of the flagellate *Synura* will produce a strange odour like that of a ripe cucumber. Algae also cause problems by blocking sand filters at water-treatment works. Blocked filters have to be taken out of commission to be cleaned or backwashed. Whilst in normal circumstances a filter might be used for between 30 and 100 hours before cleaning, an algal bloom can reduce its useful life to less than 10 hours. Diatoms cause the worst nuisance because their resistant silica walls rapidly block the interstices of the filter.

Quite apart from these operational problems, toxins released from algal blooms in reservoirs can cause outbreaks of gastroenteritis in human consumers, can kill fish in the reservoir and poison farm animals which drink the water (Prescott, 1969). Fish mortality can also occur when the algal mass decays and uses up oxygen in the water. Some of the circumstances favouring bloom formation are now well understood. Blooms occur naturally in lakes which lie on soft, easily-weathered rocks and therefore have a good supply of nutrients. They are also likely to develop in any lake which has become enriched as a result of human interference. Important sources of enrichment are municipal sewage, organic industrial effluents, drainage from cultivated land and from animal rearing units. The critical transformation from a nutrient-poor to a nutrient-rich lake can sometimes be pinpointed with some accuracy. In Lake Zurich, in Switzerland, the critical changes occurred in the 1890's when urban and industrial growth around the shores accelerated. At this time blooms of the diatom *Tabellaria fenestrata* and the blue-green alga *Oscillatoria rubescens* first appeared and massive algal growths persisted in the lake up until the 1960's when remedial measures were introduced. It has been interesting to notice in this case that the partially isolated upper basin of the lake, which has experienced little shore development, escaped those changes.

The enrichment process and the biological changes which accompany it are usually termed 'eutrophication', and the distinction is made between a nutrient-rich or 'eutrophic' lake and a nutrient-poor or 'oligotrophic' lake. The degree of eutrophication of a lake is usually gauged from the amount of plant growth in its surface waters and the degree of oxygen depletion in its lower levels. In eutrophic lakes large quantities of dead plant and animal material sink towards the bottom and are broken down by bacterial decomposition, using up oxygen in the process. The water in deeper lakes is usually divided in summer into two layers, an upper layer or epilimnion, and a lower layer or hypolimnion with little mixing between the two (Fig. 5.4). This stratification is based on differences in water density caused by the heating effect of the sun on the lake's surface layers. It effectively prevents fresh supplies of oxygen from reaching the lower levels. The oxygen-depleted zone can spread upwards from the lake bottom until it affects the entire hypolimnion.

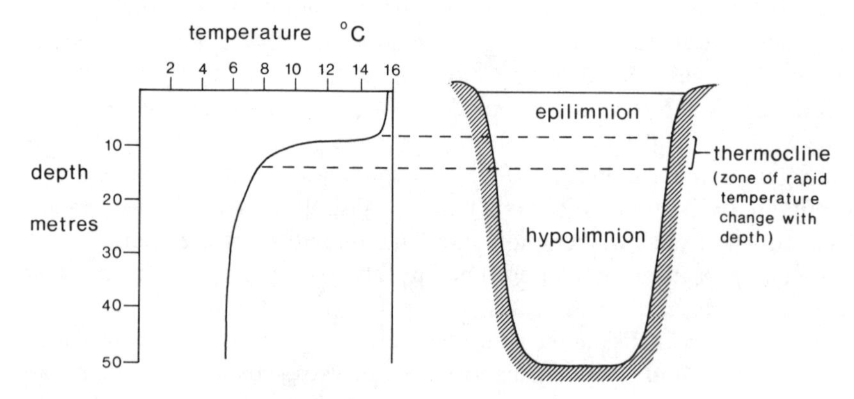


*Fig. 5.4* Summer stratification in a temperate lake

Under natural conditions it is still not clear which nutrients determine whether a lake will be oligotrophic or eutrophic. However, in most situations where eutrophication is attributable to human intervention, phosphate appears to be the key nutrient. There is a general relationship between the biological properties of lakes and the amounts of phosphate which they receive. Eutrophic lakes usually have a high phosphate input relative to their depth and oligotrophic lakes a small one. It appears that depth is one of the important factors in eutrophication and that deeper lakes can cope with greater inputs of phosphate before showing signs of eutrophication (Fig. 5.5). Confirmation of the central role of phosphate has been obtained in experiments involving the addition of nutrients to oligotrophic lakes. Phosphate additions were found to be sufficient in themselves to produce symptoms of eutrophication, including algal blooms (Schlindler & Fee, 1974).

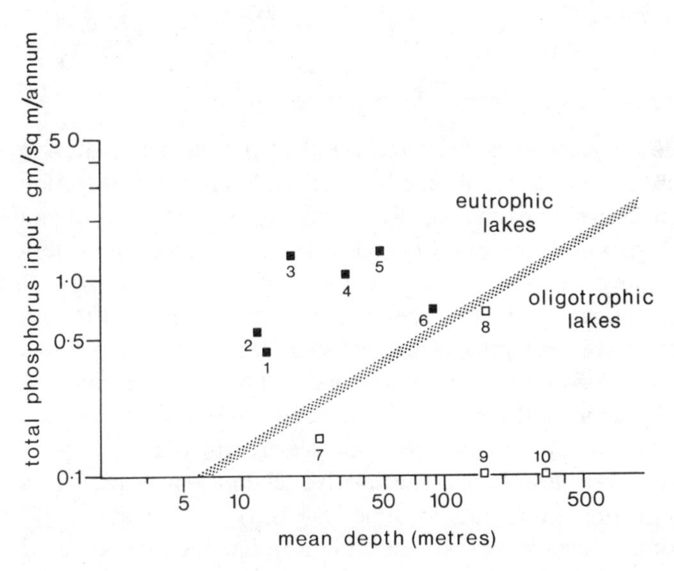


*Fig. 5.5* The effect of phosphate input and depth on lake eutrophication. Shallow lakes with high phosphate inputs are most likely to show the symptoms of eutrophication (after Vallentyne, 1974).

*1.* Furesø (Denmark)
*2.* Lake Mendota (U.S.A.)
*3.* Lake Erie (Canada – U.S.A.)
*4.* Lake Washington (U.S.A.)
*5.* Lake Zürich (Switzerland)
*6.* Lake Ontario (Canada – U.S.A.)
*7.* Lake Vänern (Sweden)
*8.* Lake Geneva (Switzerland)
*9.* Lake Superior (Canada – U.S.A.)
*10.* Lake Tahoe (U.S.A.)

Although eutrophication and bloom formation has been studied most in temperate regions, there is every reason to believe that the same general principles apply in the tropics. One of the best examples is Lake McIlwaine, a man-made reservoir which provides the main water supply for the city of Salisbury in Rhodesia. It receives processed sewage effluents from urban areas and its waters contain high concentrations of plant nutrients such as phosphates and nitrates. Algal blooms have appeared in the reservoir with increasing frequency and have created serious difficulties in the water-treatment plant (Munro, 1966; Marshall & Falconer, 1973).

Algal problems can sometimes be aggravated by a scarcity of filter-feeding planktonic animals. A good example from Britain was provided by the blooms which appeared in 1968 on Chew Valley Lake, a Bristol Waterworks reservoir. These were on an unprecented scale and rendered the water unfit for domestic consumption. A painstaking investigation showed that pollution by pesticide residues (dieldrin, lindane, DDT) from sheep-dipping had so depleted the populations of small crustaceans in the lake that they were no longer able to exert any controlling effect on the algae (Bays, 1971).

### *The control of algal blooms and water-weeds*

The numerous problems created by algal blooms and water-weeds have stimulated a search for methods of control. The demonstration of a link between phosphates and algal blooms suggests that the obvious control method is to limit the input of phosphates to vulnerable waters. This has been achieved in some cases by the diversion of sewage outfalls. The diversion of sewage from Lake Washington has reduced the incidence of algal blooms there. Similarly in the chain of lakes associated with the town of Madison in Wisconsin, the bloom has been moved progressively downstream by altering the position of the sewage outfalls (Vallentyne, 1974). Another approach is to include in sewage-treatment processes a final phase which removes phosphates. This involves chemical precipitation using compounds of iron, aluminium, or lime (calcium oxide). This so-called tertiary treatment of sewage was pioneered in Sweden and Switzerland but is now widely used in the United States and Canada, particularly in the Great Lakes drainage basin.

It has been claimed that at least 50 per cent of the phosphate in municipal sewage in North America is derived from detergents which employ phosphates (e.g. sodium triphosphate) as an ingredient to combat water hardness. Efforts have been made to find a non-phosphate substitute for this purpose and one such compound, sodium nitrilotriacetate (NTA), has been accepted for general use in some countries (Sweden in 1967 and Canada in 1973).

Because of our incomplete understanding of the factors which promote the growth of the larger water-weeds, the approach to their control has been more symptomatic. The three common methods are the application of herbicides, mechanical cutting, and the use of biological agents (Mitchell, 1974).

Many of the important aquatic weeds can be effectively controlled with herbicides; for example, the water-hyacinth is susceptible to 2,4-D and the water-fern to diquat. Submerged weeds are less easily treated with herbicides but some success has been achieved with acrolein (acrylaldehyde). Herbicide application can have unwanted side-effects. For example, although 2,4-D is relatively harmless to fish in reservoirs, if it is allowed to drift on to crops such as cotton it can cause serious damage. Fortunately few tropical impoundments with weed problems have intensive cultivation around their margins.

Various devices are available for cutting and crushing weeds. Those which return plant material to the water have the disadvantage that the dead plants can cause deoxygenation. If an economic use could be found for aquatic weeds there would be an increased incentive for developing methods to remove them from the water. The water-fern is used in a minor way as a garden mulch around the shores of Lake Kariba and in parts in India the water-hyacinth is utilised in the production of an agricultural compost. However, the most promising possibility for utilisation would seem to be in the direct feeding of dried weeds to livestock.

Finally, there is the possibility of controlling water-weeds by biological means, using the animals which normally attack the plants in their natural habitats. A number of insects have been introduced into Africa from South America in an attempt to control the water-fern and the water-hyacinth. In general the results have been disappointing. A grasshopper (*Paulinia acuminata*) imported from Uruguay and Trinidad to Lake Kariba in an attempt to control the water-fern caused some damage to the fronds but never became sufficiently numerous to control the plant. This was possibly because the grasshoppers themselves were checked in turn by predators such as frogs and the jacana, or 'lily trotter'. The water-hyacinth also has a wide range of enemies in South America including a weevil (*Neochotina*), a mite (*Orthogalumna*) and the larvae of a moth (*Acigona*), (Mitchell & Thomas, 1972). These various species are under consideration as possible control agents in Africa and the first two were released in Zambia in 1971 in a long-term trial to test their effectiveness.

Amongst the vertebrates, some fish are claimed to be highly effective in weed clearance. The grass carp or white amur (*Ctenopharyngodon idella*), a fish originating in the large rivers of South China, is reported to be efficient in clearing rooted aquatic weeds in temperate situations. Similar claims are made for various *Tilapia* species in the tropics. One of the most interesting suggestions is that a large aquatic mammal, the manatee (*Trichechus manatus*) (Fig. 5.6), could be used on a world-wide basis for weed control. The manatee's natural habit is to graze at night on aquatic plants in shallow coastal waters and rivers. The daily intake of an adult animal is reported to be 90 kg of wet vegetation. In Guyana the animals have been used since 1885 to keep ornamental pools in the Botanic Gardens at Georgetown clear of weeds (Allsop, 1960, 1961). In experiments carried out in 1959 – 60 two

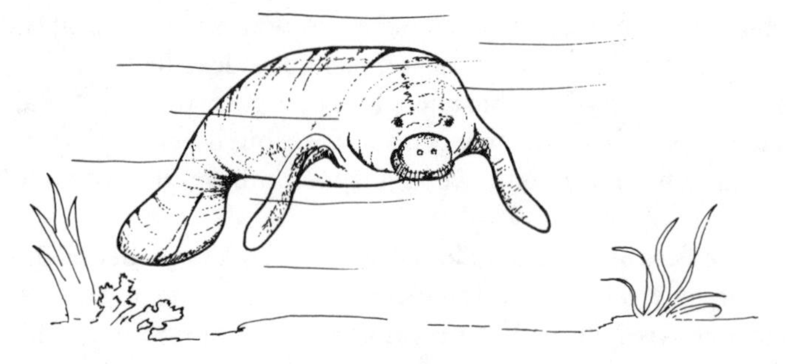

*Fig. 5.6* The manatee (*Trichechus manatus*)

manatees demonstrated that they were able to clear a weed-choked canal 7 m wide and 1450 m long. They started at one end and worked systematically to the other, completely clearing the weeds in 17 weeks. Altogether more than 120 manatees have been introduced for weed clearance into irrigation canals, reservoirs and drainage systems in Guyana. On this basis the animals would appear to have considerable promise for weed control. There are, however, some serious practical difficulties. Effective as manatees are in clearing narrow channels, they tend to be more selective in open water and will pass thick stands of one weed species to feed on another. It seems doubtful too whether enough animals could be concentrated in the problem areas of large reservoirs. They are not in any case common, nor are they easy to breed in captivity. Because of their docility and the excellence of their meat, they are also very vulnerable to human predation.

## Health implications of man-made lakes

The construction of an artificial lake especially in the tropics can have important consequences for human health. Initially the concentration of the large labour force needed to build the dam, possibly as many as 10 000 men, creates favourable conditions for the spread of communicable diseases. In addition, the eventual substitution of a large stretch of static water for a flowing river alters the capacity of the area to support the intermediate hosts of a number of important diseases.

Schistosomiasis (Bilharzia) is one of the more important diseases whose transmission is favoured by the creation of standing bodies of water. Stagnant water provides a suitable habitat for the snails which act as intermediate hosts for the parasitic worms (species of *Schistosoma*) causing the disease. Schistosomiasis occurs in Africa, Asia, the West Indies and South America and a number of different species of the parasite and of snail hosts are involved. In all the areas however the life cycle of the parasite is substantially the same (Fig. 5.7). The adult parasites lodge themselves either

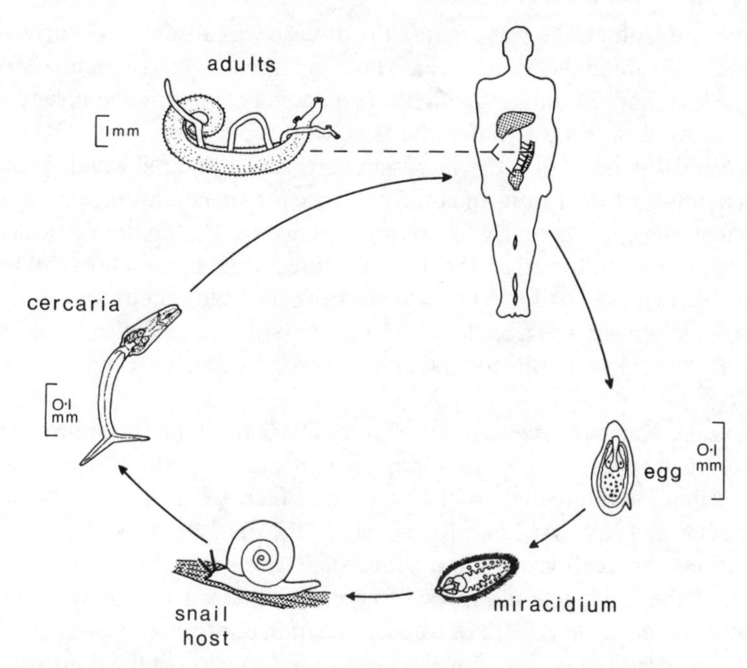


*Fig. 5.7* The life cycle of the parasite *Schistosoma*, responsible for Bilharzia or Schistosomiasis in man

in the blood vessels of the bladder (*Schistosoma haematobium*) or in the mesenteric vessels of the liver (*Schistosoma mansoni* and *S. japonicum*). The eggs shed by the parasite pass eventually into the bladder or the lumen of the gut and are passed out in urine or faeces. If they reach water, the eggs hatch and liberate a free-living larva, the miracidium. Development proceeds if the miracidium locates a snail of the appropriate species and manages to penetrate its tissues. After further developmental stages and multiplication in the snail, another free-swimming stage the cercaria, is produced. This is the infective stage for man and it enters the human body by penetrating the skin. As the water on the skin evaporates the cercariae force their way through the epidermis, partly by mechanical means and partly by chemical secretion, and at the end of 24 hours the parasite will have entered the blood stream prior to reaching the appropriate site in the body for adult development. As a single snail can emit 2000 cercariae daily and continue to do so for 200 days the risks of bathing in snail-infested waters are considerable. Infection can occur as a result of any activity associated with water: bathing, washing or collecting water for drinking.

Most of the major reservoir schemes in Africa have resulted in an increased incidence of schistosomiasis in people living around the shores.

Before the Volta Lake was created the disease was absent or occurred only sporadically along the Volta river. The two largest towns (Krachi and Yeji) had an incidence of only 1% and 3% respectively and a village survey along the east bank showed an infection level of 0·5 per cent. As the lake filled between 1964 and 1967 the increasing growth of aquatic weeds provided both a suitable substratum and food resource for an increasing population of the snail hosts, particularly for *Bulinus truncatus*. An epidemic of urinary schistosomiasis followed and by 1968 the infection rates in schoolchildren at some local villages had risen to 100 per cent. The settlements most affected were those on the west bank where the shelving shore and reduced wave action were most liable to favour the spread of snails (Paperna, 1969, 1970).

At Lake Kariba, development of aquatic weeds along the shore line has favoured the expansion of snail populations including the host species for both urinary and intestinal schistosomiasis which are now present in abundance (Hira, 1969). At Lake Kainji where the dam was closed in 1968 the conditions for weed growth and hence snails are generally less favourable. Much of the shore line is exposed to considerable wave action, with steep banks allowing little growth of weed. No detailed information was available on the incidence of schistosomiasis before construction of the dam although infected snails had been found in tributary streams in the area. However from 1970 to 1972 there was an increase in the incidence of urinary schistosomiasis from 43·9 to 60 per cent in two villages at the western end of the ferry crossing. At New Bussa, at the southern end of the lake, an increased incidence of new infections in the 10–14 year age group has been reported. At Kainji the critical areas for infection appear to be the small bathing sites established where inundated roads give access to the lake shore (Waddy, 1975).

The construction of the Aswan High Dam (completed in 1969) and the creation of Lake Nasser in the Nile valley is also likely to increase the incidence of schistosomiasis. The new arrangements for continuously irrigating about 300 000 hectares of land in Upper Egypt, whilst greatly improving the prospects for crop cultivation, will increase the numbers of disease-carrying snails. Previously their survival had been limited by the seasonal drying-out of watercourses (Farid, 1975).

In theory the situation in a snail-infested water can be improved in a variety of ways: by removing the weeds on which the snails sit and feed; by poisoning the snails using a chemical molluscicide (e.g. copper sulphate or sodium pentachlorophenate) or by improving the sanitary habits of the local population. Limited foci of infection are obviously easier to eliminate than more extensive ones but none of the suggested methods seems very promising for large artificial lakes.

Mosquitoes are another group of disease-carrying organisms which are associated with stagnant water sites. There are several instances where the creation of mosquito breeding sites by the construction of reservoirs is known to have increased the incidence of disease. In the Tennessee Valley project in the United States the building of the Hales Bar dam in 1912 and the Falling Water River dam in 1927 was followed by serious epidemics of malaria affecting people who lived within mosquito flight-range of the impounded waters. The Tennessee Valley Authority solved this problem by fluctuating the water levels of the lakes and keeping the verges free from weeds thus making conditions unsuitable for the malaria carrier *Anopheles quadrimaculatus*.

In the tropics the incidence of mosquito-borne diseases is often so high at the outset of a scheme that the creation of extra breeding sites has very little effect. However one or two species of the genus *Mansonia* are especially favoured by lake construction. The genus is important because it is implicated in the transmission of a range of virus and filarid worm infections in Africa, Asia, the Pacific and northern Australia. The immature stages of other mosquitoes obtain their oxygen directly from the air by means of respiratory siphons which pass through the film at the water surface. By contrast the larvae and pupae of *Mansonia* have respiratory siphons which are especially strengthened for piercing the submerged air-bearing tissues of aquatic plants (Fig. 5.8). These mosquitoes appear to have special affinities for the water-weeds which thrive in newly formed lakes. The impact on the health of local communities has yet to be evaluated in detail.

Onchocerciasis or river blindness is another disease which needs to be considered in relation to man-made lakes. It occurs throughout a large area of tropical Africa and affects about 30 million people. There are other, smaller, foci of the disease in South America and the Yemen. The causal agent is the filarid worm *Onchocerca volvulus* and the symptoms of the disease include dermatitis, swelling of the lower limbs (elephantiasis), the formation of nodules containing the adult parasites on the head, trunk or legs, and finally and most importantly, blindness. The blindness results from the cornea of the eye becoming opaque as the result of cellular reaction to the presence of the parasite's embryos.

The parasite is transmitted from person to person by biting flies of the genus *Simulium*. As they take their blood meal the flies pick up the minute infective stages or microfilariae from the surface layers of the skin (Fig. 5.9).

The main vector species in Africa is *Simulium damnosum*, a small fly which spends its larval and pupal period attached to rocks or vegetation in fast-flowing streams. In some of the other species, for example those in the *Simulium neavei* group in East Africa, the later larvae and the pupae often

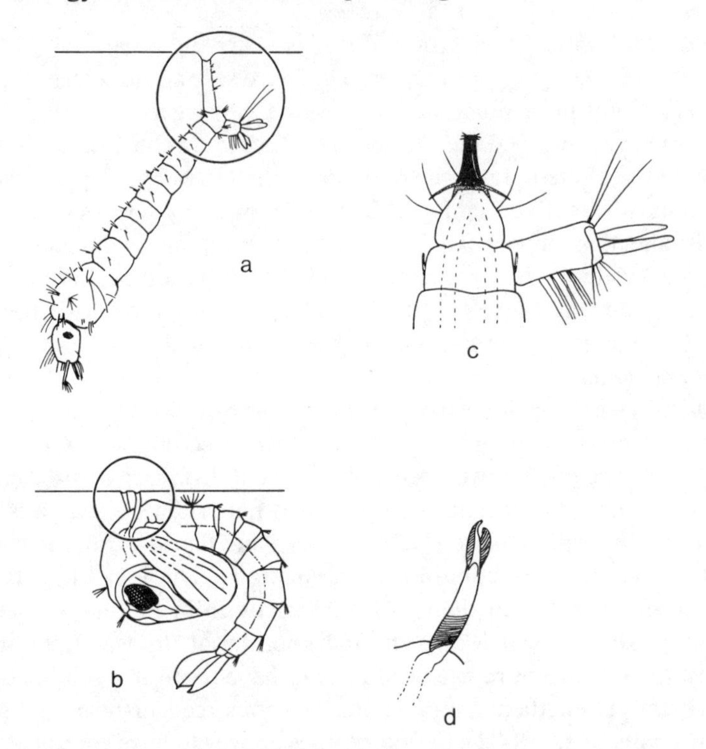


*Fig. 5.8* The respiratory siphons of a typical culicine mosquito larvae and pupae (*a, b*) compared with the specially modified siphons of *Mansonia* (*c, d*)

attach themselves to fresh-water crabs. This habit may protect them from being dislodged when floods disturb the unstable parts of the stream bed.

It would be expected that the damming of fast-flowing rivers to form lakes would reduce the incidence of the disease by eliminating breeding sites. This happened when the construction of the Owen Falls dam on the outflow of Lake Victoria eliminated a series of cascades and rapids. More usually, however, *Simulium* survives in tributary streams and is also liable to colonise the fast-flowing water on the dam spillway. Infected *Simulium* have a flight-range of many kilometres so that in the absence of special control measures the risk to human populations established around a new lake may be at least as great as when they lived alongside a fly-infested river. The usual solution is to dose the overflow water and the tributary streams with insecticides. DDT has been extensively used for this purpose and kills *Simulium* larvae at concentrations which are sufficiently low to have a minimal effect on fish and on humans. However, to exceed the worm's life span in its human host, application of the insecticides will need to be continued for 20 years. Concern has been expressed over the accumulation

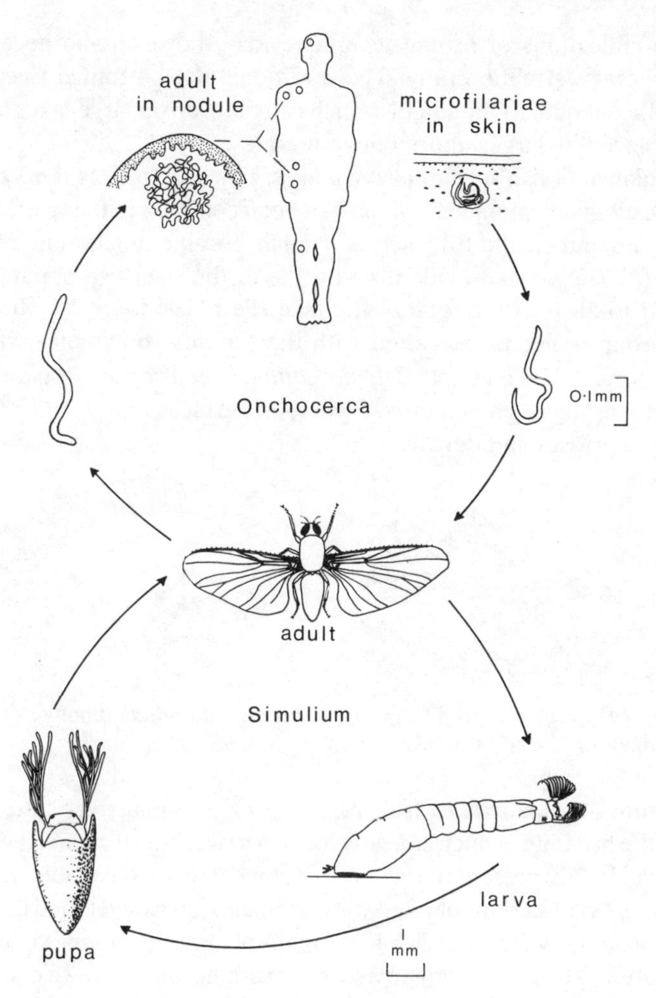


*Fig. 5.9* The life cycle of the parasite, *Onchocerca volvulus*, and its vector, *Simulium*. *Onchocerca* causes onchocerciasis or river-blindness in man

of DDT which could result from such a programme and the possibility of using alternative compounds such as abate and methoxychlor is currently being explored.

## Lake fisheries

Although new lakes may involve the loss of agricultural land, this can be offset to some degree by their capacity to support flourishing fisheries. On the north bank of Lake Kariba more than 2000 local fishermen were landing 4000 tons of fish per annum within five years of the reservoir's formation.

Some modifications of fishing technique may, however, be necessary to exploit the new situation fully. The traditional flat-bottomed river canoes are usually unequal to the squalls which can blow up on large lakes and have to be replaced by larger, more buoyant craft.

The balance of fish species also changes. For example, as the Volta lake filled, the elephant snout fish, which had been common in the river, virtually disappeared and cichlid fish such as *Tilapia* became much more abundant (Fig. 5.10). *Tilapia* had made up only 1% of the river fish population but increased to 50% of the total in the lake (Petr, 1968, 1971). The success of this group seems to be linked with their ability to exploit a variety of plant materials. For example *Tilapia galilaea* feeds on phytoplankton; *T. nilotica* takes algae growing on submerged surfaces and *T. zillii* feeds on submerged grasses and detritus.

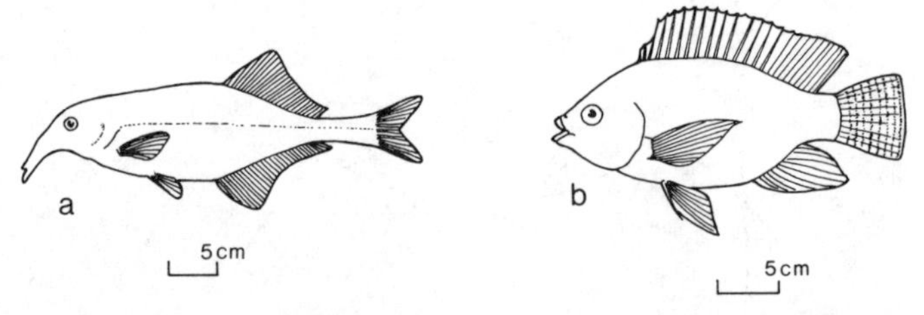


*Fig. 5.10* (*a*) Elephant snout fish (*Gnathonemus tamandua*) (family Mormyridae); (*b*) Cichlid fish (*Tilapia nilotica*) (family Cichlidae)

Sometimes the paucity of fish species in the original river system sets a limit on the fish fauna which can develop in a new lake. At Lake Kariba none of the fish already present in the River Zambezi seemed capable of exploiting the dense populations of planktonic animals which developed in the lake. It was known however that fish feeding on planktonic animals occurred in other systems. In Lake Tanganyika for example there are two species of 'sardine', *Limnothrissa miodon* (Fig. 5.11) and *Stolothrissa tanganicae*, which feed in this way and form the basis for an important commercial fishery. It was decided in 1965 to transfer 'sardines' from Lake Tanganyika to Lake Kariba in the hope of starting a fishery there also. During 1967 and 1968 the Zambian Government flew one-third of a million fry to Kariba and released them in the lake. In less than a year they had dispersed so widely

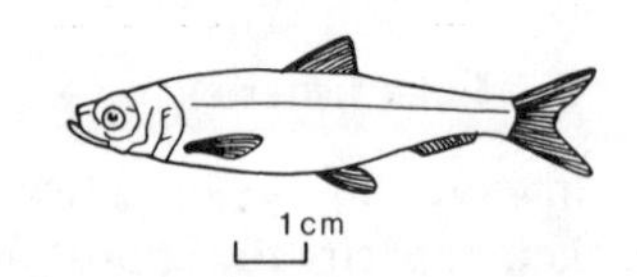


*Fig. 5.11* A Lake Tanganyika 'sardine' (*Limnothrissa miodon*) (family Clupeidae)

and successfully that they now form the basis of an important fishery. Although this particular venture was so successful, all such proposals to introduce species into new habitats need to be examined very carefully. If the receiving habitat already contains a highly evolved fish fauna, the introduction of alien species may produce no benefits for fishery interests and may disrupt settled communities of great scientific interest.

It has been recognised for some time that the preparatory work carried out on a reservoir basin can have a bearing on its future use as a fishery. When the first large lakes were being created in tropical forest areas it was considered that the trees in the area to be flooded should be felled to prevent them from interfering with subsequent fishing operations. This was done at Kariba in an operation costing several million pounds, but not at Volta or Kainji. We now have good reasons for believing that it is better from a fisheries point of view to leave the trees rather than fell them. Even at Kariba, fishermen have found it more profitable to fish in the few sites where the trees have been left. Detailed studies at Kariba and Volta have shown the reason for this. (McClachlan, 1970, 1974; Petr, 1970, 1971). The submerged tree trunks and branches form very suitable surfaces for the growth of encrusting algae and are also colonised by a variety of invertebrate animals. Notable amongst these are the larvae of the mayfly *Povilla adusta* which tunnel in the wood, feeding mainly on encrusting algae but also exploiting microscopic planktonic forms. Figure 5.12 illustrates how the

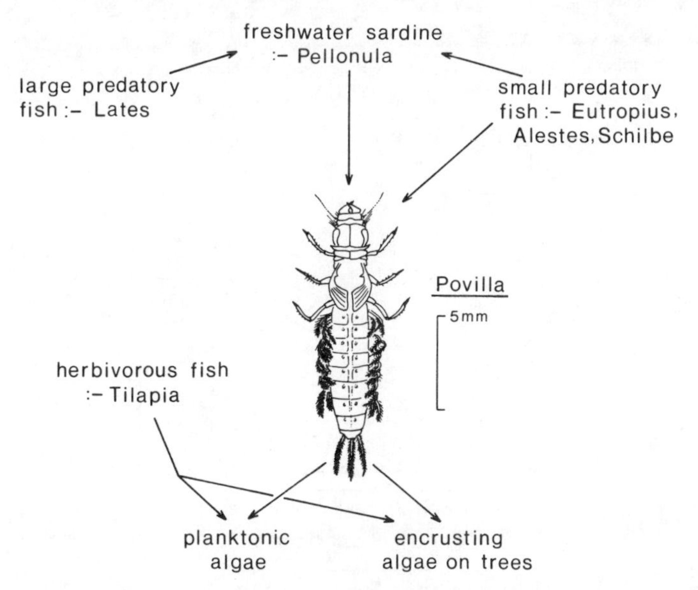


*Fig. 5.12* The role of the mayfly larva *Povilla* in the food chain of Lake Volta (after Petr, 1971)

mayfly larvae and the encrusting algae on submerged trees play a central role in the lake food chain. Herbivorous fish such as *Tilapia* use the encrusting algae directly whilst carnivorous species feed on *Povilla* both *in situ* on the trees and when it is swimming up to the surface to emerge as an adult. Eventually the submerged trees will decay and the overall productivity of the lake will then presumably fall.

Fisheries biologists are interested in any measures which increase the productivity of artificial lakes and the deliberate enrichment of the water by fertilisers has long been an accepted technique in fish culture, However, it is now apparent that a point can be reached where increasing enrichment (whether intentional or unintentional) produces unwelcome side effects. The appearance of algal blooms and an increase in the proportion of 'coarse' fish are two of the problems which can be expected. With increasing eutrophication in temperate lakes and reservoirs, 'fine' fish (salmonids and

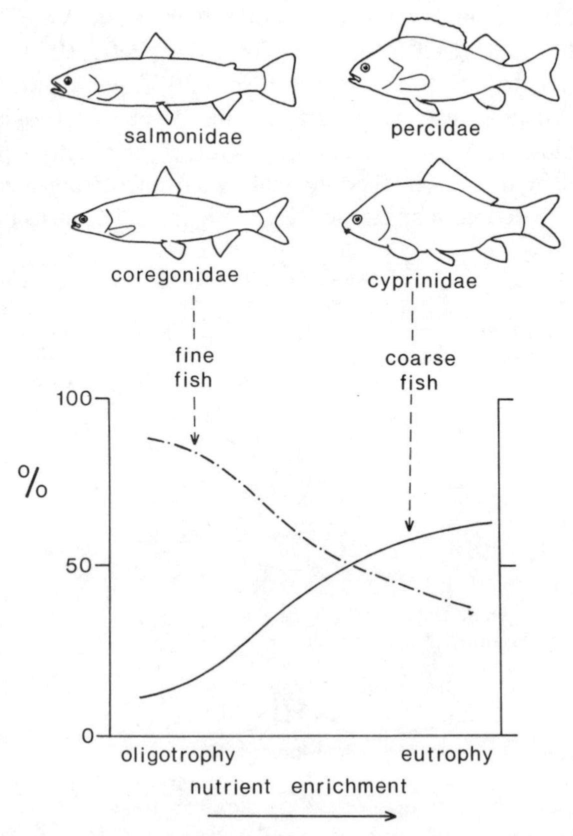


*Fig. 5.13* Some typical changes in the balance of fine and coarse fish species with increasing nutrient enrichment

coregonids) are likely to decline and 'coarse' fish (perchids and cyprinids) to increase (Larkin & Northcote, 1969) (Fig. 5.13). Coarse fish are well adapted to tolerate the depleted oxygen conditions of eutrophic lakes and to feed and spawn in their silty, weed-choked waters (Varley, 1967). To maintain trout in such eutrophic waters for sporting purposes it is often necessary to stock regularly with hatchery-reared fish to compensate for the absence of natural spawning sites. It may also be necessary to suppress coarse fish to protect the trout from undue competition.

**Migratory fish**

Dams represent a potential obstacle to migratory fish and since the beginning of the century special provisions have been made at dams in temperate areas to facilitate the movement of salmon (Clay, 1961).

There are two geographical groups of salmon (Fig. 5.14). Six species of *Oncorhynchus*, with overlapping ranges, spawn in the rivers draining into

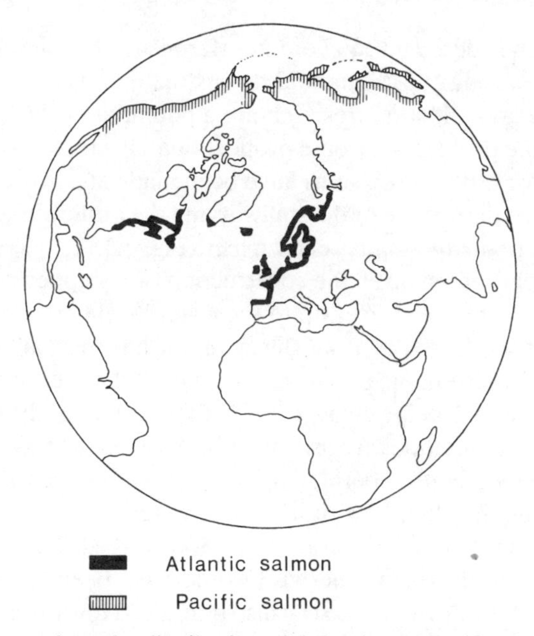


*Fig. 5.14* Map to show the distribution of Atlantic and Pacific salmon (after Jones, 1959). The six species of Pacific salmon are as follows:

| | |
|---|---|
| *Oncorhynchus masu* | masu |
| *O. tschawytscha* | chinook, spring, king, quinnat, tyee |
| *O. nerka* | sockeye, red, blueback, quinault |
| *O. kisutch* | coho, silver |
| *O. gorbuscha* | pink, humpback |
| *O. keta* | chum, dog |

the North Pacific. In the Atlantic, all the salmon belong to a single species *Salmo salar*. Some Atlantic salmon spawn in the rivers of Eastern North America, others in European rivers. Both groups of salmon spend much of their adult life at sea and move up rivers to spawn in autumn or winter. The eggs are laid on gravel beds (redds) in lakes or the upper reaches of rivers. The young salmon vary in the time they remain in freshwater after hatching, some of the Pacific species (pink and chum) move downstream in their first spring when they are only about 5 cm long. Young of the Atlantic salmon and some of the Pacific species (coho and sockeye) stay for some years in the river, migrating downstream as somewhat larger fish or smolts 6–17 cm long. In the sea, salmon feed and grow rapidly and are usually ready to return to the rivers as mature spawning fish after two or more winters. Whilst some Atlantic salmon survive to return to the river a second or third time, Pacific salmon make only a single journey and die in freshwater.

*The design of fishways*

Any dam which is designed to avoid interference with fish migration must cater for these various upstream and downstream movements. Even dams less than three and a half metres high are a potential obstacle to upstream migrants and the problems become proportionately greater as the height of the dam increases. If no provisions have been made at a hydroelectric dam, fish are attracted to the base of the spillway and the outflow from the turbine channels and are unable to proceed further (Fig. 5.15).

The usual solution involves the construction of a stepped series of pools rising from the level of the river up to the water level behind the dam. This device, variously referred to as a fishway or a fish ladder, allows the fish to move successively from one pool to the next until they reach the top of the dam. To avoid the fish being delayed whilst they search for the bottom of the fish ladder, various methods are used to guide them in the right direction. At dams with independently operated spillway gates, the centre gate can be opened wide and the others, on either side, progressively less. This produces a curved standing wave which directs fish towards the fish ladders on either side (Fig. 5.15). Fish which otherwise would have been brought to a halt along the face of the power house can also be enticed towards a fish ladder by building a channel into the power-house wall. The discharge of water down the channel and out through a series of apertures provides the necessary attraction (Fig. 5.15). This technique works even in situations like the Bonneville Dam on the Columbia River in North America where the power house is 240 m long. The top of the fish ladder is usually extended some distance upstream beyond the influence of the dam so that fish emerging are not swept back over the spillway or into the turbine tunnels.

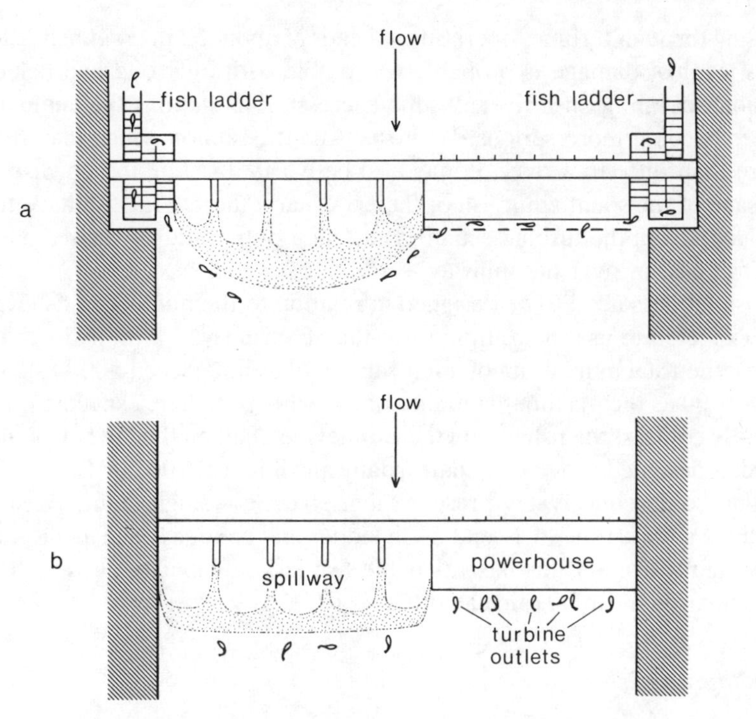


*Fig. 5.15* Plan views of hydroelectric dams (*a*) with and (*b*) without provisions for migratory fish (partly after Clay, 1961)

All these provisions are based on the knowledge that salmon migrating upstream seek out and swim up against strong currents (of the order of 1·25–2·5 m/s). At any point where fish are to be attracted, the channel dimensions need to be adjusted to produce flows within this range.

Provisions for the downstream movement of juveniles are designed on a different basis. This movement is largely passive and the young fish have little capacity to avoid strong currents. If the turbine intakes behind the dam are screened to exclude young fish, the intake channels need to be designed to reduce flow rates below about 0·3 m/s, otherwise the fish are likely to be dashed against the screens and damaged. Fish successfully excluded from the turbine channels have a chance of locating the top of the fish pass or being carried over the spillway if it is operating. An alternative strategy is to allow young fish to pass without delay through the turbines and to accept that a certain proportion will be killed. In fact, the results are not as catastrophic as might be imagined. The turbines rotate comparatively slowly and the spaces within the machines are sufficiently large to allow many smolt and fry to pass through unscathed. Tests on Atlantic salmon smolt in Scotland suggest that 80–90 per cent are likely to escape damage when passing through turbines operating at heads of about 52 m (Pyefinch, 1966). This level of damage is probably acceptable with only one hydroelectric station operating on a river. With a succession of stations the cumulative losses become more serious. With the Atlantic salmon which may return more than once to a river, it may also be worth assisting the downstream passage of the spent adult fish or 'kelts'. Clearly these large fish have to be excluded from the turbine channels and their only route downstream is via the fish pass or over the spillway.

Fish ladders need to be designed in relation to the number of fish which are expected to use them. In most of the Scottish rivers with hydroelectric dams, the total annual run of adult salmon does not exceed 5000 fish, with 500–1000 as the maximum number of fish which could be expected on one day. By contrast the fishway on the Bonneville Dam on the lower Columbia River is designed to accommodate a daily maximum of 100 000 fish of which 10 000 could conceivably present themselves in a single hour. Each fish requires a finite time to ascend a fish ladder and enough space to allow free movement; it is on this basis that the fish ladder capacity needed at any particular dam can be calculated (Clay, 1961).

### *Fish lifts and fish traps*

On many Scottish and Irish rivers where the total fish run is small by North American standards, a simpler and cheaper kind of device, the fish lift, has been used with success. It is essentially a tube with water flowing through, stretching the height of the dam (Fig. 5.16). The fish, attracted to the base of the lift by the water flow, are trapped in the lift by closing the bottom gates. The shaft then fills with water carrying the fish to the upper level where they

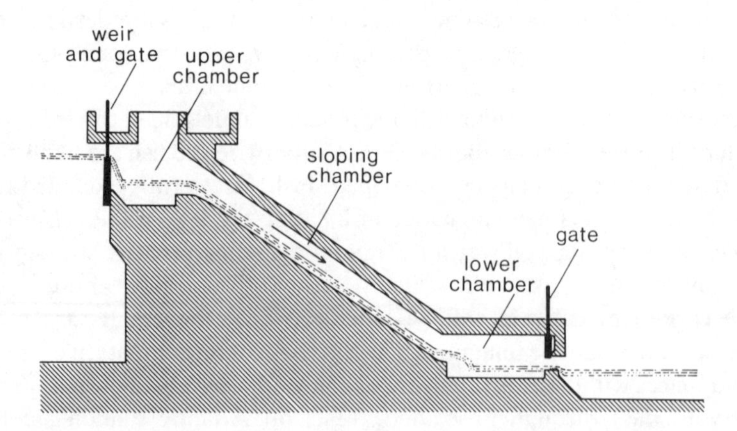


Fig. 5.16 A fish lift

are released. The inducement necessary to persuade the fish to leave the top chamber is provided by opening a by-pass valve which produces a limited flow through the system. After the fish have had time to leave the top chamber, the tube is emptied of water and the cycle starts again.

A third technique is employed on some rivers where the small runs of salmon do not justify the cost of building either a fish ladder or a fish lift. This involves trapping the migrating fish and transporting them around the dam in water-filled lorries. Most trapping systems consist of a weir or steel-rack barrier where the water flow is so arranged to attract the fish into lateral holding ponds. They can then be transferred to a tank lorry or trailer using nets or a hopper. Finally there are some rivers with so many man-made obstacles that attempts to perpetuate normal salmon runs have been abandoned. In these situations arrangements can be made to obtain eggs and sperm from trapped fish and to rear smolt in a hatchery. The smolt are then released downstream of the barriers.

*Nitrogen supersaturation below dams*

An unexpected problem which has recently come to light in connection with large dams is the development of nitrogen supersaturation in the spillway water (Beiningen & Ebel, 1970). This condition arises when water, passing over the spillway, traps air which is forced into solution as the water enters the plunge basin. When fish which have been exposed to supersaturated waters move into more normal conditions, nitrogen in the blood comes out of solution producing bubbles. These lodge in the blood vessels and tissues, and produce a condition analogous to the 'bends' developed by divers when they surface too rapidly. Serious mortality of both adult and juvenile fish in the Columbia River has been attributed to this 'gas bubble disease'. In the Snake River, a tributary of the Columbia River, the problem became serious after water was impounded at the lower Monumental Dam in 1969 and the Little Goose Dam in 1970. Nitrogen supersaturation in the river then rose as high as 146 per cent and about 30 per cent of migrating chinook salmon, both adults and juveniles, were affected.

*Fish passage at estuarine barrages*

Fishways and fish lifts can also be built on estuarine barrages. However, if any of the proposed barrage schemes are implemented some special problems are likely to arise because of the length of the dam walls. For example, one barrage proposal for the Bay of Mont St. Michel on the Normandy coast would involve a wall 42 km long and the largest of the Severn barrage schemes would have a wall of 22 km. Fish would have considerable difficulty in locating fish ladders under these circumstances. Only where large amounts of water were discharged down the fish ladder or where it was sited alongside a spillway or turbine channel would there be a good chance of attracting fish to the right point along the wall. In an estuarine impoundment designed for power generation, there would be the possibility of allowing fish to pass through the slow-moving turbines rather than trying to attract them to a fish ladder.

### *The effects of regulated flows on salmon migration*

As well as creating barriers, dam construction can affect migratory salmon by changing the flow regime downstream of the dam. Instead of the original alternation of spate and drought conditions, the flow becomes more uniform. The overall effect of this is a matter for discussion. On the credit side, regulated flows can allow fish to pass natural obstacles previously impassable at low water; on the other hand they can eliminate the sudden floods or 'freshets' which normally stimulate upstream movement. Migrations usually start during floods (Davidson *et al.*, 1943; Alabaster, 1970), although eventually fish will move upstream without this stimulus. An obvious approach to this problem is to arrange for artificial floods or 'freshets' to be released from the dam where this can be done without wasting too much water. These artificial floods have been shown to be effective in stimulating migration in some cases (Hayes, 1953).

Regulated flows can have an adverse effect on spawning grounds downstream of the dam. Spawning salmon select loosely-packed gravels through which there is the good flow of water essential for the proper development of the eggs. Natural floods in rivers serve to keep these spawning gravels free of fine sediments which would otherwise block the interstices and interfere with water circulation. Regulated flows are less likely to fulfil this function and this is a further argument in favour of the release of artificial freshets.

### *Problems associated with the transfer of water between catchments*

It is now well established that salmon which survive their period in the sea almost invariably return to the river in which they developed as juveniles and normally very few fish stray into other rivers (Harden-Jones, 1968). The fish appear to recognise their home river from chemical substances dissolved in the water. There is evidence to suggest that some of these substances are secreted by young fish developing in the river (Solomon, 1973).

In many parts of the world water-supply systems are being made more flexible by providing for the transfer of water from one river catchment to another. This raises the intriguing possibility that the release of these 'mixed' waters will confuse incoming fish searching for the characteristic odour of their own river. As long ago as 1880 Buckland had understood the homing mechanism in salmon when he wrote 'a salmon coming up from the sea into the Bristol Channel would get a smell of water meeting him. "I am a Wye salmon" he would say to himself. "This is not the Wye water, it's the wrong tap, it's the Usk, I must go a few miles further on".'

One hypothesis is that as a young salmon moves downstream it learns in succession the odour of its gravel bed, its tributary river, its main river and its estuary, and that these memories are stored in a series to be used in the reverse order when the fish returns home as an adult (Harden-Jones, 1968). In South Wales, for example, the Graig Goch reservoir scheme will include facilities to transfer water from the River Wye both to the River Severn and the River Usk (Fig. 5.17). The modern equivalent of Buckland's salmon might in future find it more difficult to decide whether the river it encounters at Newport is the Usk or the Wye. Presumably the salmon would not be confused if the mixing regimes were the same at the time of the adults' return as at the periods when young fish were migrating downstream. The possibility of confusion would also be reduced by water being stored in reservoirs where it would lose its characteristic odour.

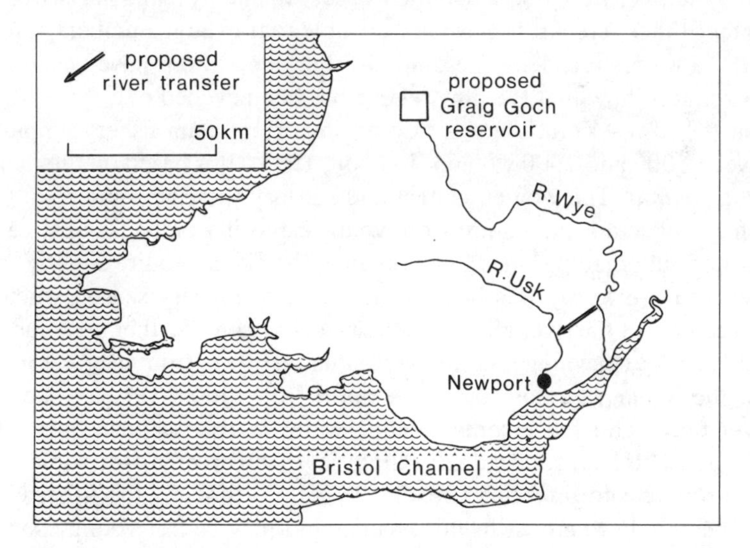


*Fig. 5.17* The proposed diversion of water from the River Wye to the River Usk

*Provisions for other migratory fish*

No other group of fish has received the same attention as migratory salmon and the only parallel is in Eastern Europe where the requirements of migrating sturgeon (*Acipenser*) have been recognised in the design of dams. The lavish provisions which are made to facilitate the migration of salmon can be justified by their importance in commercial and recreational fisheries. The world catch of migratory salmon can amount to 400 000 metric tons each year.

In the tropics there are few important fisheries based on migratory fish. Migratory eels are exploited in some African rivers and in India a fishery is based on the migratory hilsa (*Hilsa ilisha*) but these are fairly exceptional cases. Usually the fisheries opportunities created by new tropical impoundments far outweigh any interference which might be caused to migratory species; consequently, it is unusual for fishway facilities to be incorporated into tropical dams.

### *Flood fisheries*

Any type of fishery which depends on the seasonal flooding of a river is potentially vulnerable to changes in flow regime brought about by dam construction. On the River Volta for example the dam has had a marked effect on downstream fisheries. Previously 400 or 500 creeks in the lower part of the river were flooded in the wet season and by fixing nets across the mouths of these creeks, fishermen were able to trap large numbers of fish as the flood waters receded. These flood-fishery operations have been drastically reduced now that the flow regime has been evened out.

On the lower Volta there is also an important clam fishery employing between 1000 and 2000 women (Lawson, 1963). It is based on the mollusc *Egeria radiata*. The women collect and sun-dry the clams and also 'farm' them by collecting and planting out young clams in sandbanks upstream of their normal breeding sites. For breeding, the clams require slightly saline water and previously this became available in the dry season when the reduced flow of the river allowed tidal waters to move further upstream. Up until 1963 the spawning grounds were situated 30 km from the river mouth. Now the Volta dam has come into normal operation, very low flows no longer occur and the incoming saline water is able to push only 10 km upstream. This has produced an entirely new clam breeding site some 20 km away from the original one (Beadle, 1974). Fortunately the fisherwomen have been able to adjust to this change and now collect young clams for 'planting out' from the new site.

## Conservation

### *Changes at the reservoir site*

Inevitably the flooding of a river valley behind a dam can affect sites of conservation interest. In Britain a situation of this kind arose in connection with the proposal to construct a reservoir in the region of Widdybank Fell in Upper Teesdale, on the site which has become known as Cow Green. The special interest of the area is that it contains fragments of an ancient vegetation type, widespread in Britain and North-West Europe in late Glacial times, but now largely obliterated by forest growth and the spread of acidic bogs (Pigott, 1956; Godwin and Walters, 1967). The Teesdale flora probably owes its survival to the combination of severe climatic conditions and alkaline soils associated with limestone formations. Although there are similar plant refuges elsewhere in Britain, including areas on the Lizard Peninsula in Cornwall, the Burren region in western Ireland and Ben Lawers in Scotland, Teesdale is of special interest because it supports species which occur nowhere else in Britain. These include the Teesdale violet (*Viola rupestris*) and the bog-sandwort (*Minuartia stricta*), both of which have an arctic and alpine distribution in the rest of Europe (Fig. 5.18).

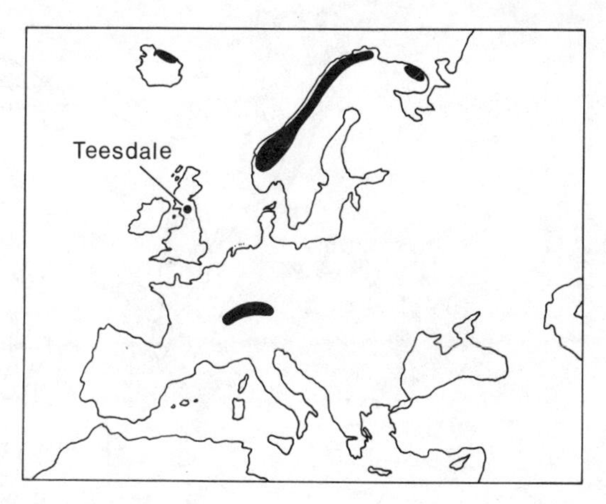


*Fig. 5.18* The European distribution of the bog-sandwort (*Minuartia stricta*)

The conservation case against reservoir construction was based on the fact that it would flood about a fifth of the limestone area supporting the special plant communities and that a proportion of the rest would be adversely affected by local climatic amelioration. This climatic effect of reservoirs has been demonstrated for a few sites where long-term temperature records are available. For example, the construction of a reservoir on the Nysa Klodzka river in Poland resulted in a local annual temperature increase of 0·7°C. However this reservoir has an area almost 20 times greater than the one at Cow Green. The case for constructing the Teesdale reservoir was based on the water requirements of expanding chemical and steel industries at Teesmouth. It was argued that industrial growth would be checked unless more water was available for abstraction from the lower reaches of the river and that consequently there was a need for a regulating reservoir on the headwaters. After a public inquiry, the water supply case was given priority. The scheme went ahead in 1967 with the anticipated losses due to flooding. It is still too early to judge whether there have been any floral changes attributable to alterations in local climate.

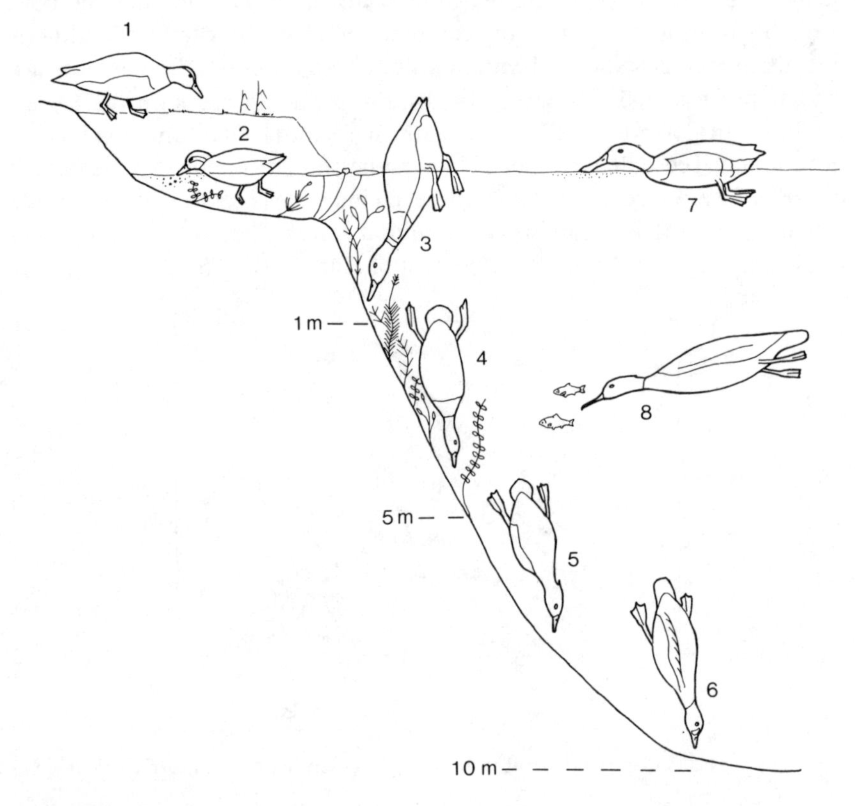


*Fig. 5.19* The preferred feeding sites of ducks on eutrophic lowland reservoirs and lakes

*1* Wigeon (*Anas penelope*)
*2.* Teal (*Anas crecca*)
*3.* Mallard (*Anas platyrhynchos*)
*4.* Pochard (*Aythya ferina*)
*5.* Tufted duck (*Aythya fuligula*)
*6.* Goldeneye (*Bucephala clangula*)
*7.* Shoveler (*Anas clypeata*)
*8.* Goosander (*Mergus merganser*)

Reservoir construction is not necessarily prejudicial to conservation interests. In Britain and elsewhere many lowland reservoirs provide important overwintering sites for wildfowl. London's reservoirs for example, which extend over 1400 hectares, regularly accommodate 7000 ducks including a wide range of species.

This diversity is related to the wide variety of feeding sites offered by eutrophic reservoirs. Each species has a favoured method of feeding and it is possible to work out a sequence of species extending from the shore into the deepest water. Such a sequence is shown in Fig. 5.19. The favoured feeding site for wigeon is on the grass of the shore, especially in areas which are free from human disturbance (Owen, 1973). Teal usually wade or swim in the shallow water near the shore and feed on seeds shed by plants at the lake margin (Olney, 1963*a*). Mallard, pochard, tufted duck and goldeneye,

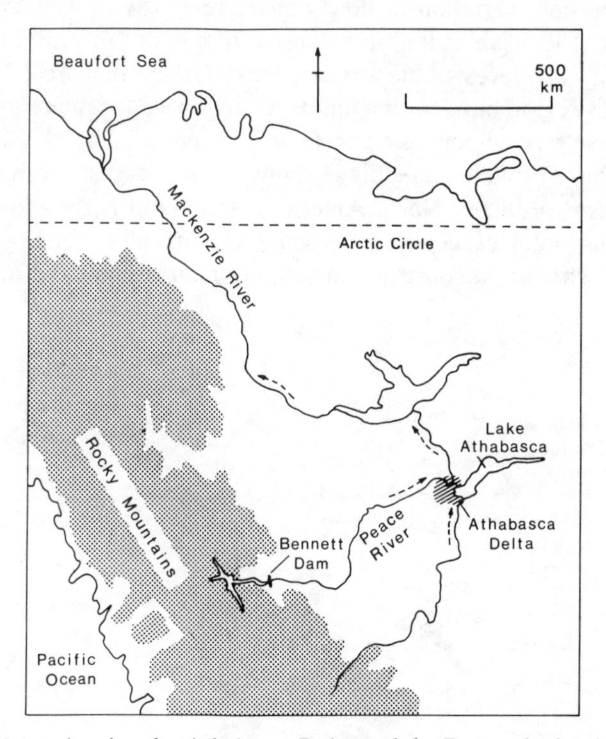


*Fig. 5.20* Map showing the Athabasca Delta and the Bennet hydroelectric dam on the Peace River

form a sequence with each species feeding in progressively deeper water (Olney, 1963*b*, 1968; Olney & Mills, 1963; Nilsson, 1972). The first two feed at depths where there is sufficient light to allow plant growth and take mainly plant food. The other two take mainly animal food from deeper levels. The shoveler has a special method of feeding, it swims along with its bill at the water surface, filtering out plant fragments and small swimming animals. Finally, the goosander is a fish-eating duck which often pursues shoals of coarse fish.

*Wildlife in flood plains*

A number of animal species make use of river flood plains and have become attuned to natural patterns of seasonal flooding. These species can be affected by the alteration of flow regimes which accompany dam construction.

The Bennett hydroelectric scheme on the Peace River in British Columbia provides a striking example of this kind of effect (Fig. 5.20). In the process of harnessing the river's spring floods, the dam has virtually eliminated the marked seasonal variation in flow which previously ranged from 140 to 8680 $m^3/s$. This has resulted in extensive drying out of a wetland area of great biological interest at the western end of Lake Athabasca. The shallow lakes in this region have been a major resting place for migrating wildfowl including the rare whooping crane(*Grus americana*). The associated sedge meadows have provided a feeding ground for the largest surviving herd of buffalo (*Bison bison*) in North America. Drying out is therefore likely to have serious biological consequences and it is difficult to see how these can be avoided without the construction of a major barrage in Lake Athabasca.

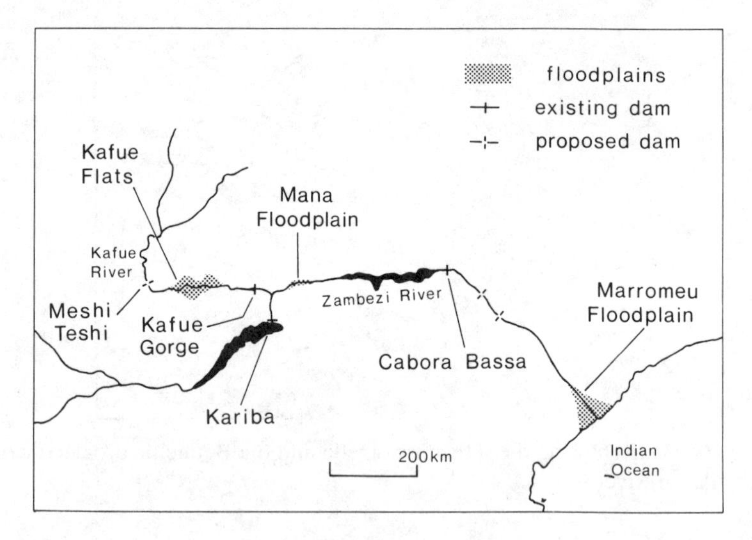


*Fig. 5.21* Map showing dams and flood plains on the Zambezi and Kafue Rivers

Adverse downstream effects have also been predicted with other hydroelectric schemes. The Zambezi/Kafue River system is of particular interest. Dams have already been completed at Kariba, Cabora Bassa and in the Kafue Gorge, and constructions at other sites are being considered (Fig. 5.21). In the Zambezi, the main concern is for wild herbivores grazing on the Mana flood plain downstream from Kariba and the large buffalo herds

(*Syncerus caffer*) living in the Marromeu area beyond Cabora Bassa. Atwell (1970) claims that Kariba (completed in 1958) has had an adverse effect on flood-plain grazing because the regulated flows have deposited less nutrient-rich silt and because, with a shorter period of inundation, the animals have been able to extend their annual grazing period and in consequence are damaging the vegetation. However, these effects are difficult to evaluate and some observers have taken the opposite view that the overall impact of the dam has been beneficial.

There are similar difficulties in predicting the effect of further dam developments along the Kafue River. The Kafue flats are notable for supporting the largest known population of wattled cranes (*Grus carunculatus*) and enormous herds of lechwe (*Kobus leche*), an antelope specially adapted to feed on flooded land. The proposal to build a storage reservoir upstream of the flats, to complement the existing dam in the Kafue gorge, has brought mixed reactions from biologists. Some observers claim that this further use of the Kafue system for power generation will be damaging to flood-plain species (Douthwaite, 1974); others argue that the two dams can be operated without producing any appreciable biological changes (White, 1973). One is forced to the conclusion that a deeper understanding of the ecology of flood plains is necessary before the downstream effects of impoundment schemes can be predicted with any accuracy.

# 6 Transport systems

## Interference from plants and animals

Transport systems, like reservoirs, can be invaded by unwelcome plants and animals. In fact some of the aquatic plants which choke reservoirs also hamper boat movements on canals and navigable rivers. With road and rail systems the main biological problems relate to encroachment by vegetation and the possibility of collisions with animals. Aircraft are particularly vulnerable to in-flight collisions with birds and this 'bird-strike' problem is now becoming recognised as an important consideration in airport siting and management.

Sometimes it is possible to predict that the siting of a transport facility or its juxtaposition to some other enterprise is likely to aggravate one of these problems. The issue then becomes significant in a planning context and here two examples are examined in detail. One involves the claim that power stations sited near harbours are likely to increase timber-damage and fouling problems. The other involves the contention that some of the operational problems of airports are a result of insufficient attention being given to the possibility of bird strikes.

### *The discharge of heated effluents to harbours*

Ships are vulnerable to colonisation by 'fouling organisms', an array of plants and animals which attach themselves to the hull below the water-line. These have the unwelcome effect of slowing the ship's progress through the water by increasing its frictional resistance. This results in increased fuel costs, and periodic dry-docking for hull cleaning. Wooden ships have the same fouling problems as steel-hulled ships but face additional hazards from wood-boring animals such as the shipworm (*Teredo*) and the gribble (*Limnoria*) (Fig. 6.1*a,b*). These species also attack wooden harbour structures, particularly pier piles and jetties. Ship-owners have known for a long time that ships are likely to accumulate a greater weight of fouling material in tropical ports than in temperate ones and it is usually reckoned that the

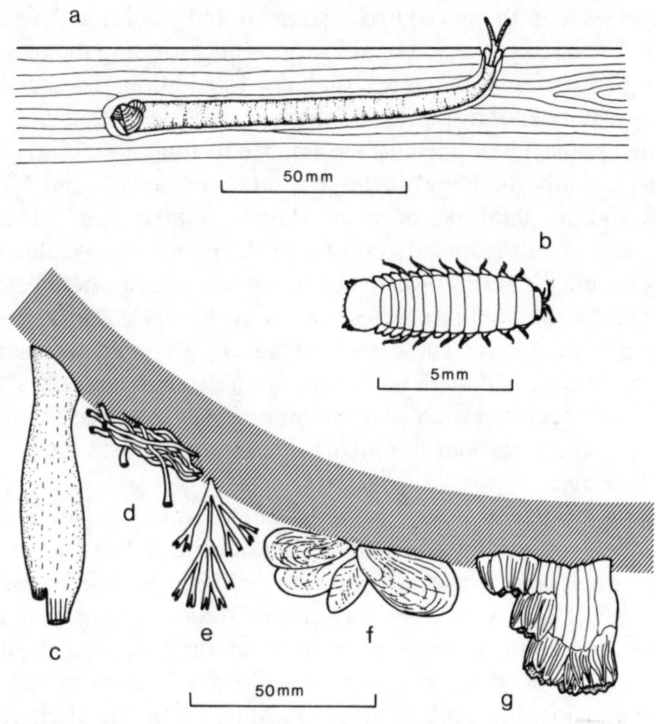


*Fig. 6.1 Some important fouling and wood-boring animals*

(*a*) Shipworm (*Teredo*)
(*b*) Gribble (*Limnoria*)
(*c*) Sea squirt (*Ascidiella*)
(*d*) Tubeworm (*Hydroides*)
(*e*) Polyzoan (*Bugula*)
(*f*) Mussel (*Mytilus*)
(*g*) Barnacle (*Balanus*)

amount of fouling is doubled (Pyefinch, 1947). This difference is attributed to the fact that fouling organisms in the tropics can reproduce all the year round whereas their temperate counterparts are checked by low winter temperatures. A similar argument applies to wood-borers.

One of the advantages of siting power stations at the coast is the possibility of using well-mixed tidal waters to disperse excess heat. This operation is likely to produce a measurable degree of local heating and it has been argued that the warm conditions so produced can cause fouling and wood-damage problems on a tropical scale.

The discharge of power-station cooling water into a dock at Swansea provided an interesting case history which has been studied in detail by Naylor (1965*a*,*b*). In this instance because of the restricted tidal flow to and from the dock, the temperature elevation was considerable, often as much as 10°C. The discharges into the dock first began in the 1930's. The biological consequences have included an increase in reproductive activity of existing fouling species (e.g. the sea squirts *Ascidiella* and *Ciona*) and an apparent increase in the damage caused by shipworms necessitating the replacement of the wooden jetties by concrete structures. In addition, the dock has been invaded by a number of tropical and subtropical species presumably brought in on ships' hulls. These include the important fouling barnacle *Balanus amphitrite*, the tube-building worms *Hydroides incrustans* and *Mercierella enigmata*, and the plant-like polyzoan *Bugula neritina* (Fig. 6.1). Another invader was the sub-tropical wood-borer *Limnoria tripunctata*. Partially offsetting the effects of this invasion by warm-water forms has been the loss of certain temperate fouling species such as the barnacles *Balanus crenatus* and *Elminius modestus*. These were apparently unable to tolerate the warmer conditions. Although no figures on the economics of the situation are available it seems certain that the multi-purpose use of the dock as a cooling pond and a harbour has increased the problems of timber damage and ship fouling.

With larger capacity power stations and a greater awareness of the damage which appreciably elevated temperatures can cause to commercial fisheries, a small dock would no longer be seriously considered for cooling purposes. Nevertheless even modern power stations discharging into less constricted tidal waters may not be without their biological effects. In Southampton water where the Marchwood power station is recorded as producing a temperature rise of up to 3°C 400 m from the discharge point, (Pannell *et al.*, 1962), there is evidence to suggest that wood borers and fouling species have increased in numbers and extended their breeding period. Here it seems that the temperature rise has not been sufficiently great to eliminate troublesome temperate species. Altogether there is a good case for examining in a comprehensive way the economic consequences of discharging heated effluents into port areas.

Paradoxically a careful examination of other effluents in ports leads to the conclusion that some have beneficial rather than detrimental effects because of their biocidal properties (Nair & Saraswathy, 1971). In Valparaiso Bay, wooden piles situated alongside the discharge from a gas-processing plant were found to be less prone to shipworm attack. Shipworm activity also seems to be depressed by both oil and sewage. The harbours of New York, San Francisco and Marseilles are all said to enjoy this benefit of sewage pollution.

### *Bird strike problems and airport planning*

Two major air disasters are known to have been caused by collisions between aircraft and birds. In 1960 at Boston, U.S.A., a Lockheed Electra crashed on take-off, killing 62 persons, after colliding with a flock of starlings. In 1962 in Maryland, U.S.A., a Vickers Viscount flying at 6000 ft collided with a flock of whistling swans. Part of the tailplane broke off and the aircraft dived vertically to the ground killing all 17 persons on board. There have also been numerous incidents in which disaster has been narrowly averted. In 1962 at Turnhouse Airport, Edinburgh, a Vanguard with 76 persons on board flew into a large flock of gulls immediately after take-off. Two of the four engines failed almost immediately and the other two lost power. A crash was averted by the pilot who just managed to maintain height, circle, and make a successful landing. One hundred and twenty five dead gulls were recovered after the incident. In North America there have been at least two incidents where collisions with birds put three of the four engines on jet airliners out of action during landing and another incident where a three-engined airliner lost two engines after colliding with a flock of gulls during take-off. Fortunately, on all three occasions the aircraft managed to land safely. Modern jet aircraft have engines which are notably vulnerable to damage by birds and they fly at speeds which give birds little time to take evasive action. Because of the large carrying capacity of modern airliners (over 400 passengers in a 'jumbo' jet) the consequences of an accident could be extremely serious. As well as the possible loss of life there are the more general problems of disrupted flight schedules and the cost of repairs.

Figure 6.2 shows some of the groups of birds frequently involved in collisions with aircraft. They range from swans weighing 10 kg to starlings and dunlins weighing about half a kilogram.

Basically the answer to the problem must lie in the sphere of airport planning and management. There is relatively little scope for strengthening airframes or incorporating protective devices into engines without loss of performance. At any potential airport site, both natural and man-made features play a part in determining the pattern of bird movement.

### *Gulls and refuse tips*

Of the man-made features, urban refuse tips have a special significance. It has been claimed that 50 per cent of all bird strikes involve gulls and it has become apparent in recent years that there is a close correlation between the siting of rubbish dumps and the establishment of dangerous flight patterns by gulls. Both airports and domestic refuse tips tend to be sited on the edges of cities, sometimes in close proximity to each other. We now know that gulls commute daily between their rubbish tip feeding sites and night-time resting places. Reservoirs are particularly favoured for resting, and to site a refuse tip on the opposite side of an airport from a reservoir is to invite problems. The daily commuting journey for a gull can involve a round trip of as much as

*Fig. 6.2* Birds frequently involved in collisions with aircraft: (*a*) Swans (e.g. whistling swan, *Cygnus columbianus*); (*b*) Gulls (e.g. herring gull, *Larus argentatus*); (*c*) Geese (e.g. lesser snow goose, *Anser caerulescens*); (*d*) Waders (e.g. dunlin, *Calidris alpina*); (*e*) Songbirds (e.g. starling, *Sturnus vulgaris*)

130 km. Refuse tips therefore need to be separated from airports by an appreciable distance to achieve a high measure of flight safety. Some countries, by usage or legislation, no longer permit refuse tips within 12 km of major airports; the ideal minimum separation is probably in the region of 30 km. This does not imply that a serious hazard exists at all airports where this degree of separation is not achieved. Many airports have escaped the

problem for fortuitous reasons rather than as the result of conscious planning. At London Heathrow, 300 000 gulls roost on reservoirs within a 15 km radius of the airport but fortunately most of them by-pass it when moving to their feeding grounds.

*Other local bird movements*

Many local bird movements are less obviously related to man-made features and are more difficult to influence. These include the twice daily journeys of wading birds between tidal feeding areas and their high-tide resting places and the movements of rooks (*Corvus frugilegus*) and starlings in the vicinity of their large winter roosts. Starlings in autumn form communal roosts, often numbering many thousands of birds. These roosts are usually formed in scrub, woods or reed beds. Each morning the flocks disperse, travelling as far as 40 km to their feeding grounds, returning again in the evening. These returning flocks have been shown by radar to extend over distances sometimes as great as 8 km and clearly can represent a hazard if their flight path takes them near an airport. Rooks can also present problems when dispersing from and returning to their communal winter roosts, which may contain birds from as many as 50 rookeries. Outside the airport perimeter it is more difficult to effect the habitat modifications needed to combat these risks but there is sometimes the possibility of modifying potentially dangerous flight patterns by collaborating with landowners to destroy local roosts.

*Problems associated with long distance migration routes*

Some airport sites have the disadvantage of being on the line of bird migration routes or flyways. These routes are used annually by birds moving between their breeding and overwintering area and may be used by flocks of hundreds of thousands of birds. In the southern part of British Columbia a number of airports lie along the route used by Canada geese (*Branta canadensis*). The geese migrate in spring from the interior valleys of the Western United States to breed in the Yukon and Central Alaska. Repeated observations have shown that the bird movements can be expected on a series of days in the second half of April, that the number of birds will be something in excess of 50 000, and that they will be flying below 15 000 ft (Myres and Cannings, 1971).

A similar situation occurs each spring in the Winnipeg region and involves lesser snow geese (*Anser caerulescens*) migrating North to their breeding areas around Hudson Bay. In the spring of 1969 a collision occurred at Winnipeg International Airport between a Boeing 737 (a short-haul jet airliner) and a flock of lesser snow geese. The aircraft was seriously damaged but managed to land safely. As a result of this incident, a detailed investigation was made of the migration patterns of the geese with the aim of reducing the possibility of further collisions (Blokpoel, 1974). It had been generally known for some time that the eastern populations of lesser snow geese overwinter along the coast of the Gulf of Mexico and in spring migrate northwards to their breeding areas around Hudson Bay (Fig. 6.3). The

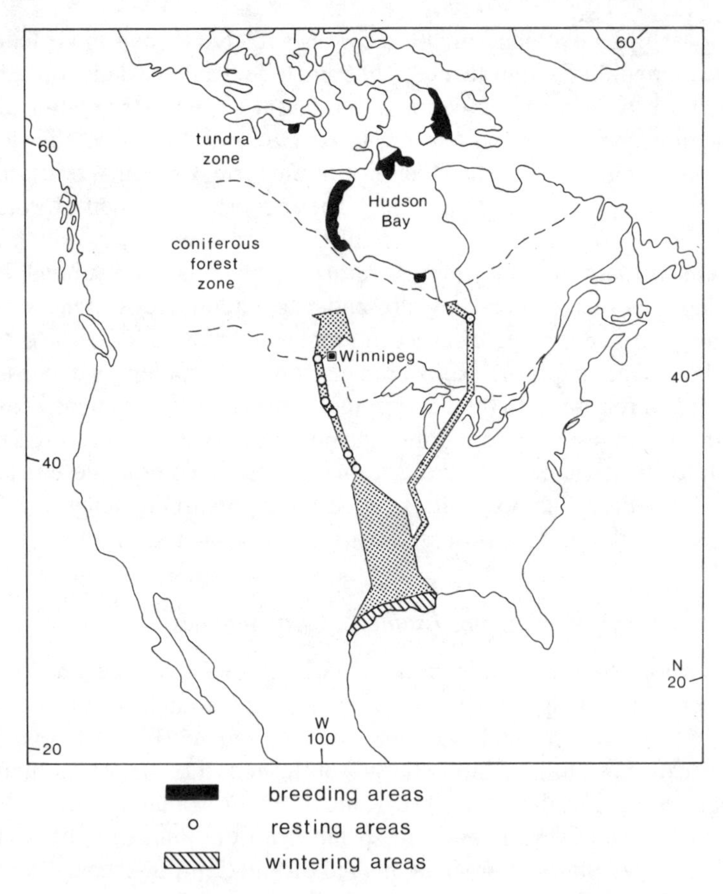


*Fig. 6.3* The spring migration routes of the lesser snow goose (*Anser caerulescens*) in North America (after Blokpoel, 1974)

journey starts in February and the geese make their way slowly across the plains following the retreating ice-line, pausing at intervals to rest on large lakes and marshes. By the end of April or the beginning of May, the birds have reached the edge of the coniferous zone and the entire flock, which may number up to 500 000 birds, becomes concentrated in a small area south west of Winnipeg. The last stage of the journey involves a dramatic departure of the flocks from this final resting site and a direct, probably non-stop, flight across the forest zone to the breeding areas on the tundra. This final breakout of the flocks is of considerable significance for air traffic movements in the Winnipeg area. It usually takes place during the first half of May and is probably related to weather conditions.

These migratory movements represent a special problem because there is no real possibility of diverting the flocks. On the other hand the hazard is of limited duration and with proper biological information its timing and pattern are predictable. Consequently the usual solution is to modify aircraft movements according to predictions and radar observations of the bird flight patterns.

*Birds attracted to airports*

Whatever bird populations are present in an area before an airport is built, the airport itself will alter the pattern and serve as an attraction for some species. This situation can give rise to a number of problems which are concerned with management rather than planning. For some birds, airports have the special attraction of including large areas of short grass. Flocks of rooks and starlings come to probe in the turf for their invertebrate foods, and gulls and waders use the areas as resting places. Resting birds tend to favour sites where their view is not restricted by tall vegetation, presumably because this makes it easier to detect approaching predators.

There is the obvious possibility of discouraging such birds by allowing the grass to grow longer. This has been tried at a number of airports and it has been found that grass longer than about 20 cm is indeed less attractive. There are however some complications. Although much of the grass cutting carried out on airports is done for reasons of general tidiness and is not strictly necessary, there are some areas which have to be kept short for operational reasons, for example, around landing lights, on grass runways used for light aircraft and in areas where there is a fire risk from jet exhausts. Alternative methods have to be used to discourage birds in these places. One possibility is to reduce the attractions for starlings and rooks by killing the soil fauna using insecticides. There are, however, the usual problems of cumulative toxicity unless short-life chemicals are used.

The long-grass strategy itself occasionally creates unwanted side effects, as at Vancouver airport where the build-up of grass-feeding rodents in the long grass attracted short-eared owls (*Asio flammeus*) which then became a potential hazard. Five hundred owls were trapped and released elsewhere over a three-year period.

Apart from the height of the grass, its species composition is sometimes important. Many commercial seed mixtures contain a proportion of white clover (*Trifolium repens*). The nitrogen-fixing properties of the species make it a suitable choice for agricultural or reclaimed industrial land. On an airfield, however, it can be a liability because it attracts wood-pigeons (*Columba palambus*). There is therefore a good case for leaving it out of airport seed mixes or eliminating it subsequently with selective weedkillers.

### *Bird scaring devices*

One method of moving birds from airfields is to scare them with disagreeable noise stimuli. Various noise-making devices have been used for this purpose (Brough, 1968). The automatic gas cannon produces loud bangs at regular intervals by exploding propane or acetylene gas; the 'shell cracker' is a modified Verey pistol, firing charges which detonate about 100 m from the pistol. Unfortunately birds become accustomed to loud noises after a time, so that their response to these devices falls off. Tape recordings of distress and alarm calls of birds are generally more effective in producing dispersal because they are linked to the birds' natural behavioural responses. Different birds respond differently to the calls, some disperse immediately (lapwings and starlings), some only after they have investigated the source of the sound (gulls and corvids). Because of this delayed response in some species, recordings have to be played at least five minutes in advance of an aircraft movement to avoid the risk of diverting the birds into its path. Birds usually respond best to the calls of their own species and next best to those of closely related species. Some birds such as wood-pigeons and oystercatchers (*Haematopus ostralegus*) either do not have distress calls or do not respond to them and the only hope with these species is that they will join the departing flocks of those birds which have responded.

Another scaring technique trades on the instinctive avoidance reaction which birds show towards birds of prey. At some airfields attempts have been made to disperse birds using radio-controlled model aircraft built to simulate a hawk silhouette: broad wings, long tail and stubby head. Live birds of prey, particularly falcons, have been used in a number of countries. Trained falcons will drive away some unwanted bird species during daylight hours but have the limitation that they will not work in fog, heavy rain or high winds. An unusual technique is used at London's Heathrow Airport. Gulls and corvids are effectively dispersed when a man standing silhouetted against the sky 100–200 m from the birds raises and lowers his extended arms at a frequency of 24 movements a minute. It is claimed that this works because the movement resembles the wing beats of an eagle.

### *Bird hazard surveys and airport planning*

There is an increasing realisation that the assessment of possible bird hazards should form part of the appraisal of potential airport sites. In the study of sites for an alternative airport for Copenhagen, one site was noted as having the 'considerable disadvantage' of being on the line of the bird migration route between south-west Sweden and Denmark. However, this was only one of the 25 factors which were considered. There has yet to be a case where the bird strike hazard caused the rejection of a site which was otherwise suitable.

It is more usual for a site to be selected on other grounds and then for a study to be commissioned to identify possible hazards and find ways of reducing them. A good example of this approach is provided by Saul's study of possible hazards at Auckland International Airport in New Zealand (Saul, 1967). The airport is sited on the edge of Manukau harbour, an area of tidal mud-flats frequented by a wide range of birds including waders, gulls, terns, swans and ducks (Fig. 6.4). As many as 12 000 wading birds feed on the mud-flats in the early months of the year. These birds created difficulties by moving between their resting and feeding areas along the original shore-line, a course which took them across the airport runway (Fig. 6.4*a*). However it proved possible to divert the birds around the end of the runway using scaring devices (*d*). Another problem arose from the movement of waders from Manukau harbour to occupy hightide resting areas in an adjacent bay. One of these 'overspill' movements took the birds across the other end of the main runways (*b*). It was concluded that this behaviour was associated with a shortage of natural resting sites around the harbour. Accordingly two artificial roosts were constructed for the birds, one at Puketutu Island where a traditional roost had been destroyed earlier by the construction of sewage ponds, and one at Wiroa Island immediately adjacent to the airport (*e*, *f*). On Wiroa Island special care was taken to make the site attractive to birds by clearing scrub and marsh vegetation to provide more open conditions. Special pools were constructed which the birds could use for bathing. By the end of the study the birds were reported to be using the new roost with increasing frequency.

The airport also suffered from gull problems and in one incident a gull ingested into the engines of a DC8 air liner caused extensive damage. Initially, gull movements in the area were largely dictated by the presence of a municipal refuse tip at the head of an inlet east of the airport. Each day hundreds of gulls moved to and from the tip crossing the eastern approaches to the runway in the process (*c*). By removing the tip the gull population in the area was immediately halved. Another major attraction for gulls was an area alongside the airport were fisherman habitually gutted their fish. To reduce the number of gulls in this area it is proposed to establish a zone (*h*) where fishing and anchoring are prohibited.

Apart from problems associated with coastal birds, the grassy areas at the airport were frequented by flocks of starlings and banded dotterels searching for food. At the time of the survey, consideration was being given to the

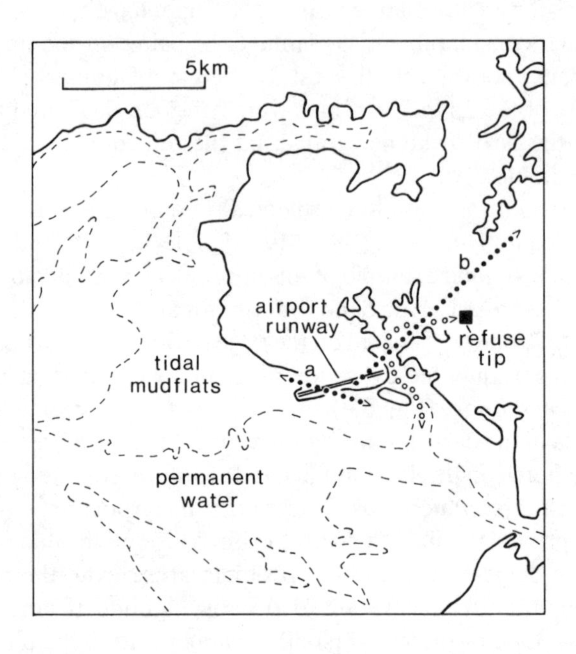


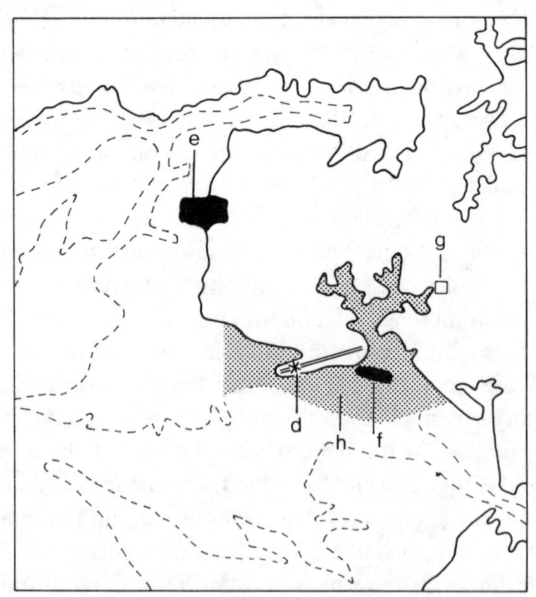


*Fig. 6.4* Bird movements in the vicinity of Auckland International Airport (*a–c*) and the measures introduced to combat bird strike hazards: (*d*) scaring devices; (*e, f*) artificial roosts at Puketutu and Wiroa Islands; (*g*) discontinued refuse tip; (*h*) prohibited fishing and anchoring zone

usual technique of allowing the grass to grow to discourage these species and also to deter resting gulls and waders.

Each airport is likely in this way to come under the influence of its own set of bird populations and the implications for safety merit consideration at an early planning stage.

## The environmental impacts of transport systems

Whilst efficient transport systems are an essential part of our everyday existence they can also be a major source of disturbance. Their construction can disrupt a wide range of urban and rural land uses, and their operation can be a source of noise and pollution. As the most rapidly expanding transport systems, road and air traffic networks yield numerous examples of the application of ecology to transport planning.

### *Air traffic systems*

Airport noise is the environmental issue which most people associate with air traffic systems. Atmospheric pollution at airports is usually of negligible importance and subsonic airliners cruising at their normal operational height create no significant problems. Supersonic airliners however are in a different category and need to be considered separately from other aircraft (p. 125).

Around any airport it is possible to draw a noise 'contour' map, which may represent the present noise situation or be a prediction for some future date based on estimates of air traffic changes. Figure 6.5 shows such a ten-year forecast map for Zurich Airport. The units used for the noise contours on these maps have been designed to reflect the degree of annoyance likely to be felt by people living within the different noise zones. Social surveys have shown that annoyance is related to two factors: (a) the total number of aircraft heard during the day and (b) the average peak noise level as each aircraft approaches and recedes from the observer (Burns, 1973). Since the degree of annoyance can be increased by either more aircraft or louder ones, the two parameters can be combined into a single function, the Noise and Number Index (NNI). Having drawn a map with NNI noise contours and having accepted the fact that there is a stepwise increase in annoyance with each contour interval, the important next step for planning purposes is to decide the level of annoyance tolerable in residential areas. The Swiss authorities, in keeping with many others, attach particular significance to the 55 NNI contour. They consider that all areas within this contour are unsuitable for residential use and that existing housing in the 45–55 NNI zone should have special protection against noise. Moreover it is felt that the

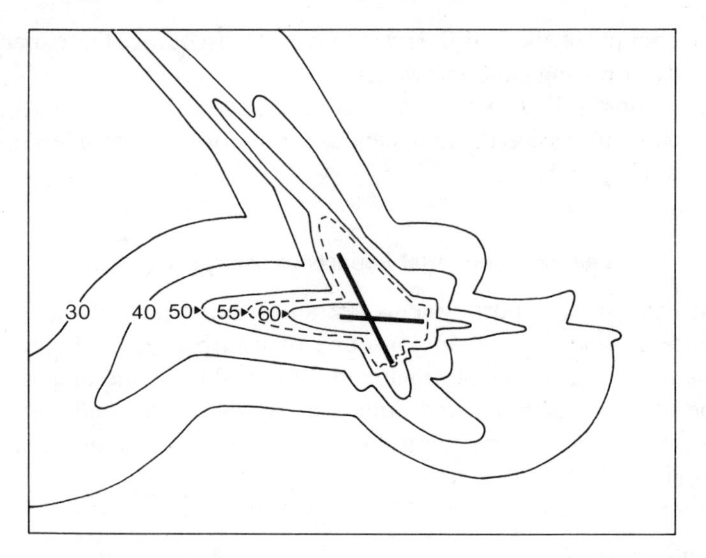


*Fig. 6.5* Noise and number index (NNI) contours in the vicinity of Zurich Airport (after Reinhard, 1975)

more stringent 45 NNI standard should be applied to new housing schemes. Not all countries use the NNI system. In the U.S.A., the Composite Noise Rating (CNR) or the Noise Exposure Forecast (NEF) are preferred. Whatever system is used however, they are all designed to predict the likely level of annoyance and to give the planner the opportunity to select an appropriate threshold level.

Sadly the introduction of noisy jet airliners in the early 1960's and the overall increase in air traffic has meant that these thresholds were exceeded long ago for many thousands of people living close to metropolitan airports. Various remedial measures have been tried. One method is to sound proof houses, as has been done around Heathrow (London), Schiphol (the Netherlands) and several German Federal airports. Whilst in moderate noise conditions this can produce an acceptable indoor environment, it does nothing to improve conditions out of doors. Another possibility is to require the airport to operate in a way which produces less noise. This can entail a curfew on night flights, the concentration of aircraft into approach and departure corridors away from residential areas or the modification of landing and take-off procedures. Ultimately, the scope for some of these operational noise abatement procedures is limited by safety considerations.

More promising are the noise improvements resulting from advances in engine design (Noise Advisory Council, 1974). The new wide-bodied jets (e.g. Boeing 747, DC10, Lockheed Tristar) are appreciably quieter weight-for-weight than the first generation of four-engined jet airliners (e.g. VC10,

Boeing 707, DC8). They have engines which have been made more efficient by reducing the turbulence of mixing between the propulsive gas jet and the outside air and this also makes the engines quieter. With the increasing use of these quiet high capacity airliners one can foresee the time when the noise contours around airports could contract even though the number of passengers increases. Although this would be by far the most satisfactory solution to the aircraft noise problem, it is unlikely to come about very quickly. Many of the older noisier aircraft, have commercial lives of fifteen years and will be operating for some time to come. Indeed, it is estimated that in 1980, 700 pre-1969 four-engined jets will still be in use. Supersonic aircraft are also likely to present a fairly intractable noise problem because, with their swept back wing configuration, they will always need high-thrust engines for take-off.

*Remote siting of airports.* The siting of airports away from urban areas appears at first sight to represent a fundamentally different approach to the airport noise problem. The proposed transfer of Copenhagen's International Airport to the offshore island of Saltholm for example could dramatically improve the noise situation over the city, without the need for any other noise abatement measures (Fig. 6.6).

Unfortunately even sites which are designed to be remote from residential areas are unlikely to remain isolated. Every new airport needs a large work force to man it (100 000 workers were forecast for Maplin), and these people must be housed and provided with facilities nearby. Most economists would also argue that a new airport should be regarded as a growth centre for commercial industrial and urban development. A site adjacent to an airport certainly provides special opportunities for trading companies and manufacturers of high value products suitable for dispatch by air. On this basis there is the risk of an originally remote site developing the typical noise problems of its metropolitan counterpart. Admittedly at a new site it should be possible to control development from the outset and squeeze residential areas between predicted noise contours. However, unexpected changes such as new approach corridors or extra runways could negate the whole strategy.

Whatever the subsequent evolution of remote airports, they are frequently characterised at the planning stage by conflicts with existing users of rural land and from conservationists. Toronto's second airport at Pickering is to be sited on grade 1 agricultural land. An airport proposed for the Severn estuary in Britain would interfere with a commercial salmon fishery and destroy a mudflat area used by wading birds (p. 150). The Maplin scheme is famous for the controversy it raised about the fate of brent-goose flocks. Conservation issues are particularly difficult, partly because of the problem of assessing the conservation value of a site in financial terms and partly

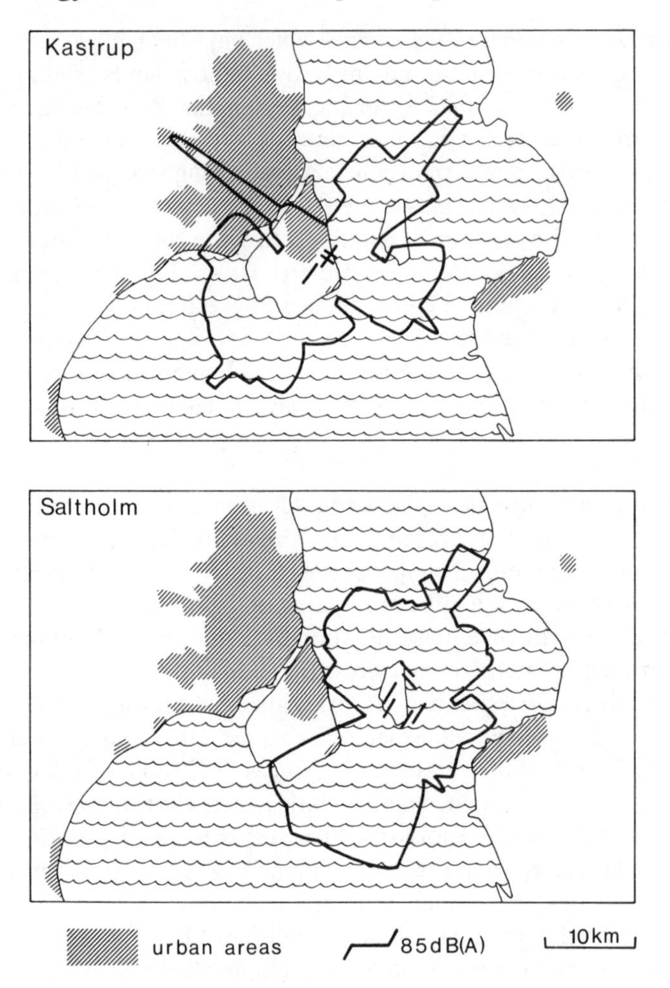


*Fig. 6.6* The effect of transferring Copenhagen's International Airport from Kastrup to Saltholm (after Lemberg, 1975)

because our biological knowledge is often insufficient to predict the consequences of a development with any accuracy. The geese at Maplin are a case in point. It was perfectly clear that the construction of the airport on a reclaimed mudflat north of the Thames estuary would destroy a large section of the most extensive bed of eelgrass (*Zostera*) in Britain (Fig. 6.7). It was also well established that these beds were visited each winter by over a third of the world population of the dark-bellied brent-goose (*Branta bernicla bernicla*). These facts taken in conjunction with the belief that eelgrass was an essential part of the birds' diet led to the then reasonable conclusion that these overwintering birds were seriously threatened (Ogilvie & Matthews,

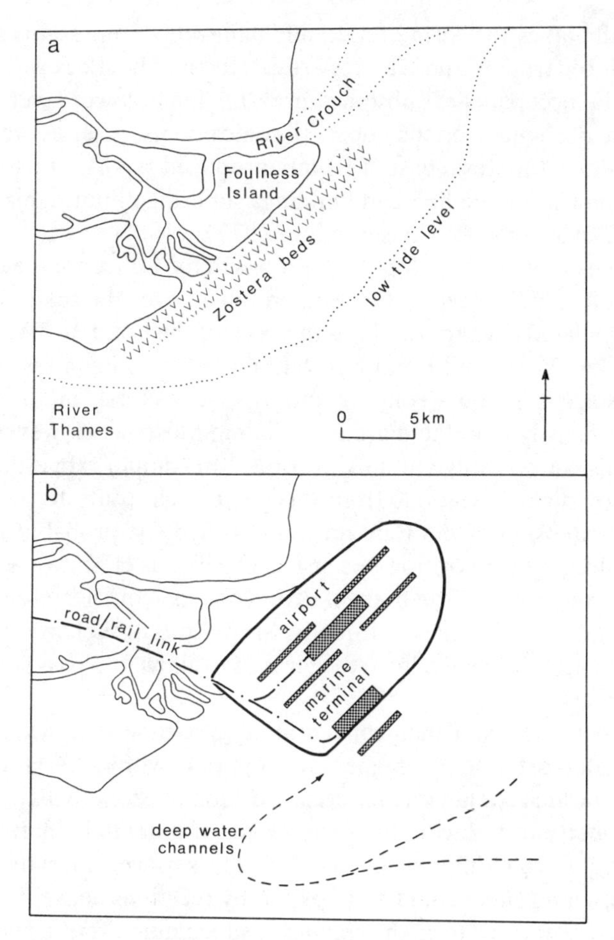


*Fig. 6.7* The Maplin development (after Howard, 1970)

1969; Drake, 1973). Since then however, the geese have flourished and have demonstrated by attacking farm crops that their dietary preferences extend well beyond eelgrass. This suggests that the impact of the airport, whilst certainly detrimental, might not have been as disastrous as had at first been imagined. As the Maplin project has since been abandoned we now have no way of knowing the answer.

*Supersonic airliners.* Supersonic airliners raise separate issues because of their great speed and their operation at high altitudes at the edge of the stratosphere. Concorde cruises at 1350 m.p.h. (compared with 600 m.p.h. for a large subsonic jet) and at this speed it pushes in front of it a pressure wave which curves down to the ground and is heard as a sonic boom. The boom usually takes the form of a double bang which represents the shock waves produced from the nose and tail respectively. The shock wave can also be felt by the occupants of subsonic aircraft flying at lower levels. Various studies have been made on the impact of sonic booms. A series of Concorde test flights down the west coast of Britain produced reports of people being startled, animals stampeding and buildings suffering minor damage (Committee for Environmental Conservation, 1972).

With some of the routes adopted or proposed for Concorde the boom will create no difficulties because it will be produced over the sea or over very sparsely populated land areas. This is the case for the London – Washington, London – New York, and London – Bahrain routes. In contrast the proposed two-stage London – Tokyo route via Novosibirsk in the U.S.S.R. passes over heavily populated areas, including Moscow. However because each stage is comfortably within Concorde's maximum range there is the possibility of slight deviations from the great circle route to avoid urban areas. The extension of the Bahrain route to Sydney, probably via Singapore, is a different proposition because each stage is close to the aircraft's range limit and there is little hope of diversions to avoid heavily-populated regions of India and Indonesia. Equally, for reasons of fuel economy, there is no real chance of slowing down the aircraft to allow it to pass silently over sensitive areas.

As well as flying faster than conventional jet airliners, Concorde and its Soviet counterpart also fly higher and operate within the zone of the stratosphere which contains a high concentration of ozone. This ozone band has the property of reducing the penetration of potentially harmful ultra-violet radiation from outer space to the earth's surface. It has been argued that the screening effect could be impaired by reactions between the ozone and nitric oxide emitted from the engines of supersonic aircraft, which would destroy a proportion of the ozone (Johnston, 1971). An increased incidence of skin cancers might be one of the consequences of more ultra-violet light reaching the earth's surface.

Any attempt to reach a conclusion about the risks involved has been hampered by a lack of information about the dynamics and chemical reactions of the atmosphere and about the relationship between ultra-violet concentrations and skin cancer incidence. Nor have the discussions been free from political overtones. The whole issue stimulated a great deal of new research in these areas. The present consensus (Valery, 1975; World Meteorological Organisation, 1976) seems to be that currently planned supersonic transport aircraft, because of their relatively low flight altitudes (17 km) and limited numbers (30 – 50 projected), are unlikely to produce any effects which could be distinguished from natural variations. It is predicted however that a large fleet of supersonic aircraft flying at greater altititudes would be a different proposition and might have a noticeable effect on the ozone layer. To prevent this eventuality, it might well be necessary to control nitric oxide emission levels by international agreement.

*Road traffic systems*

*Noise*. The problem of road traffic noise can be approached in a similar way to airport noise. In both cases there is a need to define the noise level which can be regarded as acceptable in residential areas and to define the relationship between noise production and the type and volume of traffic. Before examining the problems of road traffic noise in any detail, we need to look again at the question of sound units. Sound travels through the air as a series of pressure variations. Basically sound pressures are expressed in newtons per square metre ($N/m^3$), a unit now internationally known as the 'pascal'. The range to which the human ear responds is approximately 0·00002 to 20 pascals. For everyday purposes however, this basic scale is too unwieldy and is replaced by the decibel (dB) scale, which provides a more manageable series of numbers over the audible range (Table 6.1). This scale actually expresses the ratio between a reference level (usually 0·00002 pascal, the faintest sound that can be heard by most healthy young adults) and the pressure of the sound being measured. The scale is a logarithmic one and this means that an increase of 10 dB roughly corresponds to a doubling of loudness, thus a noise of 80 dB is twice as loud as a noise of 70 dB. Table 6.1 gives some typical examples of noise levels.

Unlike aircraft noise, road traffic noise is generated exclusively at ground level, so that special interest attaches to the capacity of screens and barriers to block transmission. The NNI system used for aircraft noise is quite unsuitable for road traffic because the sound of individual vehicles is often not distinguishable. Instead some measure is needed which reflects the frequency and intensity of noise peaks during the working day. The unit most generally used is the $L_{10}$ (18 h) index. This is an average of the noise level exceeded for 10 per cent of the time during each hour of the period from six in the morning to midnight on a normal working day. The readings are made on a noise meter with the scale weighted electronically to compensate for the response of the ear (the 'A scale') and are expressed in 'A weighted' decibels, dB(A). Unlike the NNI index, there has been no special effort to make the $L_{10}$ scale correlate with annoyance level. In some circumstances this is a major drawback.

The current recommendation in Britain is that the $L_{10}$ noise level outside a house should not exceed 68dB(A), previously the recommended level was 70 dB(A). Various indoor standards have also been suggested, these include 50 dB(A) for a house alongside a busy urban street and 45 dB(A) for a suburban house away from main traffic routes. Once these standards were

*Table 6.1* Some typical examples of noise levels in relation to the decibel and pascal scales.

| *Decibels (dB)* | *Sound pressure pascals (Pa)* | *Examples* |
|---|---|---|
| 120 | 20 | |
| | | Jet aircraft at 150 m |
| 110 | | Inside boiler-making factory<br>'Pop' music group |
| | | Motor horn at 5 m |
| 100 | 2 | |
| | | Inside tube train |
| 90 | | Busy street<br>Workshop |
| | | Small car at 7·5 m |
| 80 | 0·2 | |
| | | Noisy office |
| 70 | | Inside small car<br>Large shop |
| | | Radio set – normal volume |
| 60 | 0·02 | |
| | | Normal conversation at 1 m |
| 50 | | Urban house<br>Quiet office |
| | | Rural house |
| 40 | 0·002 | |
| | | Public library |
| 30 | | Quiet conversation<br>Rustle of paper |
| | | Whisper |
| 20 | 0·0002 | |
| | | Quiet church |
| 10 | | Still night in the country<br>Sound-proof room |
| | | Threshold of hearing |
| 0 | 0·00002 | |

defined it became obvious that they were not being achieved in many urban areas. Even in 1970 when the standard was set at the less stringent 70 dB(A), it was calculated that 8·5 million people were being subjected to noise levels greater than this. So as far as planning is concerned, the task is not only to ensure that new housing and new road schemes meet the required standard but also to improve conditions for houses already sited in unacceptable noise environments.

The problem can be approached at source by attempting to reduce the sound produced by the noisiest vehicles. Heavy lorries are the most important group in this category and there is a project under way concerned with the production of two quiet heavy vehicles with 200 b.h.p. and 350 b.h.p. engines whose noise production would be little greater than an ordinary saloon car. The use of foam plastic acoustic shielding for the engines of heavy vehicles is also being explored. Arrangements to smooth the flow of traffic in towns and divert through-traffic on to by-passes could all help to alleviate the heavy vehicle problem. However, the view that reducing traffic volume on a road will necessarily reduce noise problems, is something of an illusion engendered by the adoption of the $L_{10}$ system. At some traffic levels, a reduction in traffic flow will actually increase annoyance because the useful masking effect of background noise has been diminished. This response would not be predicted from the $L_{10}$ scale because the system takes little account of background noise. Other measures such as the Traffic Noise Index (TNI) and the Noise Pollution Level (NPL) take more account of the relationship between noise peaks and background noise (Noise Advisory Council, 1973) and so are the more appropriate for predictions about annoyance levels. Thus in the situation shown in Fig. 6.8 a reduction in vehicle numbers from 5000 to 2000 an hour has little effect on peak noise levels ($L_{10}$) but produces an appreciable reduction in background noise ($L_{90}$). The likely increase in annoyance is accurately expressed by the $L_{NP}$ curve.

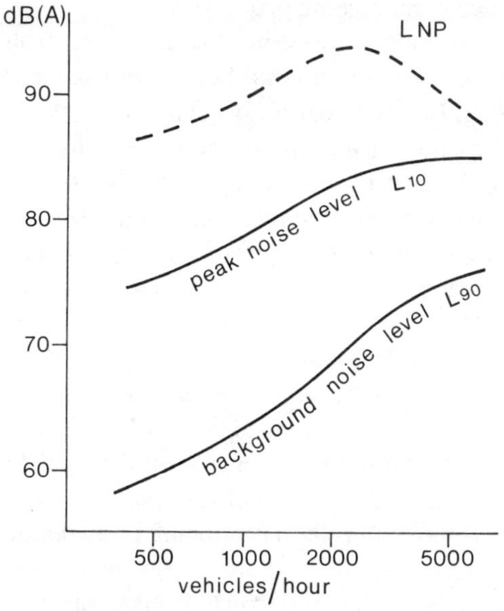


*Fig. 6.8* The relationship between traffic flow and noise level expressed on the Noise Pollution Level ($L_{NP}$) scale

The simplest methods of reducing noise disturbance are either to increase the distance between the source and the observer or to erect barriers between them. On a level site at distances greater than 15 m noise levels fall by 4 dB(A) each time the distance from the carriage-way is doubled. Barriers can take many forms: they can consist of walls or earth banks near the carriage-way or intervening buildings. The effectiveness of barriers in reducing noise levels can readily be assessed using routine calculations (Department of the Environment 1974a, 1976b).

In spite of these various possibilities for reducing noise, many urban houses still have unacceptably high external noise levels which interfere with the use of gardens. In these circumstances, the only solution may be to salvage the indoor environment by introducing sound-proofing measures such as double glazing and special ventilation systems.

*Pollution.* Although it is well established that unburnt hydrocarbons and nitrogen oxides from motor vehicle exhausts are involved in photochemical smog formation, it has also become clear that, for climatic reasons, this problem is likely to remain limited in its geographical distribution (p. 37). Much more general concern has arisen from the knowledge that lead and carbon monoxide, which in other contexts are well known as poisons, are also major constituents of exhaust fumes. In addition the polycyclic hydrocarbon, benzpyrene, a by-product of many combustion processes including those which take place in diesel engines, has been shown in animal experiments to have carcinogenic properties.

The main question must be whether the emissions from motor vehicles particularly in congested urban conditions represent a significant health hazard to the public. Lead can certainly act as a poison when taken into the body in sufficient quantities and there are many well-documented cases of poisoning amongst industrial workers and in children who have chewed materials coated with lead-rich paints. The amount of lead in the blood is a good indication of the possibility of poisoning and serious symptoms usually appear at blood levels exceeding 80 $\mu$g/100 ml, the normal range being 10–30 $\mu$g/100 ml (Chisolm, 1971). People with blood showing only slightly elevated levels between 30 $\mu$g and 50 $\mu$g/100 ml rarely show any clinical symptoms. They are, however, prone to some minor metabolic disturbances and the interpretation of these is central to the whole issue. The disturbances involve the appearance in the urine of two unusual substances, coproporphyrin and delta-aminolaevulinic acid (ALA). The two compounds are known to have an important role in producing body pigments, in particular the cytochrome pigments involved in energy production and the iron-containing pigment haemoglobin which is part of the blood's oxygen-transport mechanism. Their appearance in the urine is a sign that the production pathways are becoming impaired and that intermediate products are accumulating and being excreted. It seems probable however that the system has sufficient spare capacity to withstand this disruption, particularly where it occurs at low blood lead levels and produces no clinical symptoms. This knowledge helps to put the road traffic lead question into some perspective. Even in the most exposed groups in towns, such as traffic policemen, there is little evidence for raised levels of blood lead or urine metabolites (World Health Organisation, 1969). In the absence of even these mild signs of lead-induced disturbances in the people most at risk, it is difficult to believe that the public at large is exposed to any significant hazard.

Much the same view can be taken of carbon monoxide emissions. Carbon monoxide has the capacity to combine with haemoglobin, the oxygen-carrying pigment of the blood, and to lock it up in the useless form of carboxyhaemoglobin. Severe or prolonged exposure to the gas, with blood carboxyhaemoglobin levels above 10 per cent, causes impaired vision and as the concentration rises beyond 20 per cent, progressively more serious symptoms appear including headache and dizziness and eventually coma and death. Now the higher carbon monoxide concentrations measured in towns (about 500 $\mu g/m^3$) would only be sufficient to raise the carboxy-haemoglobin level to 3–4 per cent and this is well below the threshold where clearly recognisable symptoms appear. There are claims that subtle disturbances of perception and behaviour are produced even at this low level (Schulte, 1963; Beard & Wertheim, 1967; World Health Organisation, 1972a). However, these are very difficult to reconcile with the fact that a 4 per cent concentration is common amongst cigarette smokers who do not apparently suffer from these disturbances.

There are similar difficulties in interpreting the significance of benzpyrene emissions from motor vehicles. This compound first aroused attention when it was isolated from gasworks pitch and was shown to produce tumours when painted onto the skin of laboratory animals. Since then it has become recognised as a regular minor constituent of town air and one which is derived from various combustion processes including those which occur in diesel engines. Although there is a suggestion that benzpyrene, in conjunction with cigarette smoking, can cause lung cancer in man (Royal College of Physicians, 1970), there is nothing to suggest that the contribution of pollutants from motor vehicles is of particular significance.

*Remedial measures.* Although there is no convincing evidence for regarding motor vehicle emissions as major health hazards, there could be a case for increasing safety margins, partly because of our still incomplete understanding of the effects of low concentrations of pollutants and partly to protect the small proportion of the population who are likely to be especially vulnerable. With carbon monoxide, the people with cardiovascular disorders represent a special group. For them even a small reduction in the oxygen-carrying capacity of the blood can have serious consequences. With lead, the people who could benefit from reduced motor vehicle emissions would be those who already have a high intake from other sources.

In terms of remedial action, measures aimed at smoothing the flow of urban traffic will automatically reduce carbon monoxide emissions because vehicles which are moving freely produce much less carbon monoxide per unit length of road than those which are constantly stopping and starting. Lead is added to petrol to increase its combustion properties or octane rating. The same result could be achieved although more expensively by more selective refining. Within this framework there is scope for some reduction in lead levels with only a marginal increase in cost. There comes a point however, at about 0·40 g/1, where further decreases become very expensive.

In cities with photochemical smog problems, control is a much more urgent matter. One approach has been to develop catalytic converters for incorporation into car exhaust systems. The catalyst is used in a pellet form to give it a large surface area and is effective in removing hydrocarbons and carbon monoxide from the exhaust fumes. However most catalysts have the disadvantages that they are deactivated by leaded petrols and they are not effective in reducing nitrogen oxides. The more fundamental approach of modifying the operation of the engine also has its drawbacks, because although the use of a richer fuel mixture would reduce the emission of nitrogen oxides it would increase the output of carbon monoxide and hydrocarbons (Fig. 6.9).

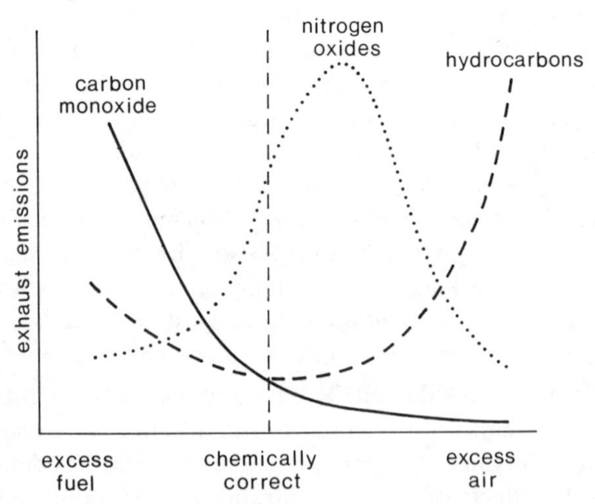


*Fig. 6.9* The relationship between air/fuel mixtures and exhaust emissions

*New Roads*. In developed areas new roads are usually justified on the grounds that, by supplementing existing overloaded roads, they will reduce journey times and cut accident rates. In undeveloped regions new roads often serve a somewhat different function and are often regarded as the instruments of agricultural and urban growth. For example, the new road systems in the Amazon basin (Fig. 6.10) are designed to support settlements at 10-km intervals and to have cultivated strips many kilometres wide laid out on either side of them.

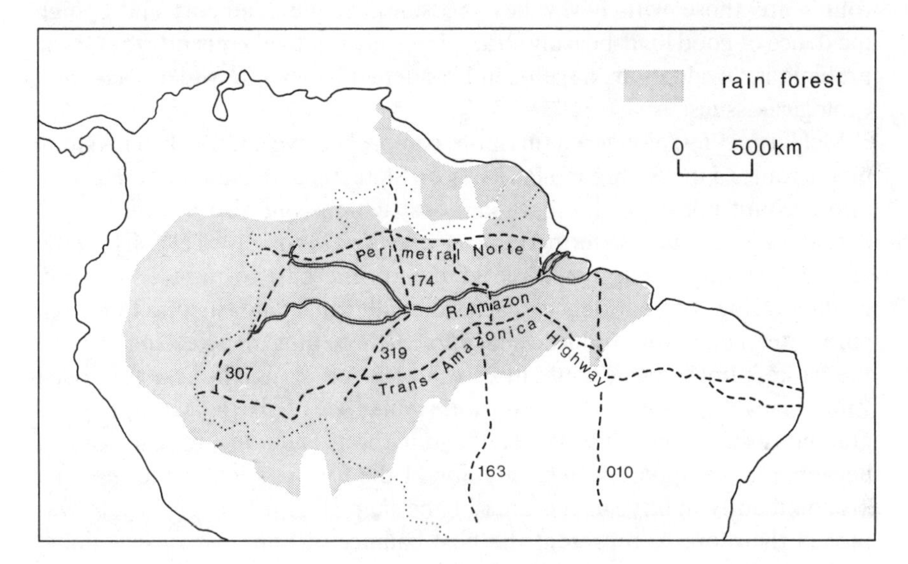


*Fig. 6.10* New roads in the Amazon basin

Whatever the circumstances, new roads are a source of disturbance. In Amazonia there are fears that the greater accessibility provided by the new roads will increase the uncontrolled exploitation of the forest fauna (Goodland & Irvin, 1975). Species particularly at risk include the giant Brazilian otter (*Pteronura brasiliensis*) hunted for its fur, and a wide range of monkey species which are exported alive to serve the international trade in pets and animals for medical research. The rare white-nosed saki (*Chiropotes albinasus*) is especially vulnerable because one of the new roads will bisect its only known refuge. Equally important will be the social, medical and cultural impacts on the Amerindian communities (Brooks, 1973; Goodland & Irvin, 1975). Of the 90 communities whose territories will be affected by the roads, at least 45 have rarely or never been exposed to outside contacts.

In the crowded rural landscapes of developed countries the pattern of disturbance is rather different and objections to new roads are often based on the disruption of farm holdings, the impairment of landscape quality and the destruction of wildlife habitats. Like remotely-sited airports, new roads in rural areas often eventually serve as focal points for urban and industrial development.

*Road alignments.* After a decision has been made to build a road from A to B the next step is usually to select the final alignment from a range of alternatives. Traditionally, engineering considerations have been the dominant factor in making this choice, and on these grounds the cheapest routes are those with few valley crossings, gentle gradients and a high incidence of good load-bearing strata. It is only relatively recently that these preliminary evaluations have been broadened to include wider social and ecological issues.

McHarg (1969) pioneered this more comprehensive approach in his study on the route of the Richmond Parkway on Staten Island, New York. He took into account not only physiographic obstructions but also a series of ten considerations which he termed 'social values'. This involved classifying the land in the study area into three grades on the basis of its scenic value, wildlife value, recreation value, residential value and so on. For each parameter a map was produced shaded with three tints of increasing density, the darkest tint indicating the most valuable zones. When these ten social value maps were super-imposed on a similar set of six indicating physiographic obstructions (the darker the tint the greater the obstruction), it became readily apparent that only one of the four possible routes lay in a predominantly lightly shaded area of the map (Fig. 6.11). This route was judged therefore to represent the best balance of minimal constructional costs and minimal interference with existing interests and was subsequently adopted. This type of exercise has much to commend it, not simply as a device to assist the planner in his decision-making, but also as a demonstration to the general public that all the important considerations have been taken into account.

## The spread of pests

Transport systems have played an important part in the accidental dissemination of plant and animal species to new geographical regions. Most of these species ride on or in some transport vehicle, but with canals it can be the channel itself which provides the dispersal route. Frequently these aliens have a disruptive effect on the settled biological communities which they invade, and many have become serious pests of agriculture and fisheries (Elton, 1958).

Many countries have carefully designed routine and emergency procedures to check the international spread of agricultural pests. These include

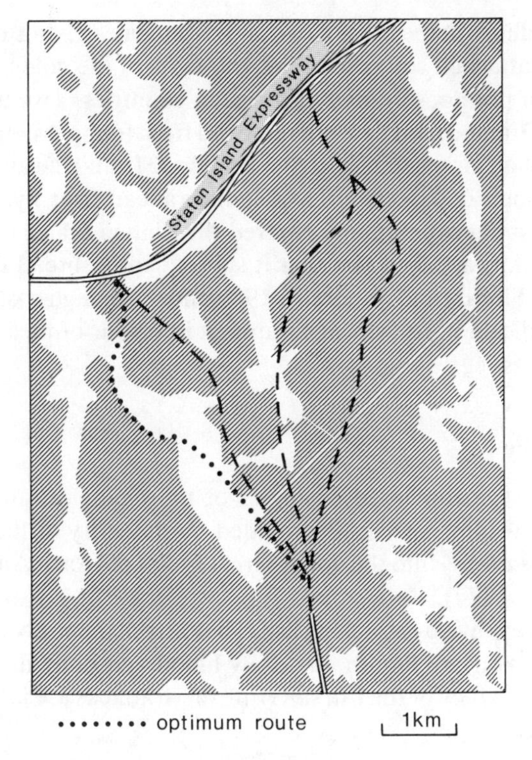


*Fig. 6.11* A simplified version of McHarg's map to determine the optimum route for the Richmond Parkway

the inspection of cargoes and travellers' baggage, the fumigation of cargo holds and the application of insecticides to the perimeters of airports and seaports. Sherman (1966) has described some of the experiences of the U.S. Department of Agriculture in controlling what he calls 'the insect jet-set'. One persistent invader from Europe has been the cockchafer beetle (*Melolontha melolontha*). This is a serious agricultural pest in Europe. The adults damage fruit trees whilst the larvae attack roots of cereal crops, beet, strawberries and potatoes. It appears from time to time in aircraft arriving in the United States and would be a most unwelcome addition to the North American fauna. To combat this and similar hazards, airport perimeters on both sides of the Atlantic are routinely sprayed with insecticides, as also are the cargo compartments of aircraft.

Sherman also describes how plant material carried in passengers' suitcases is another possible source of agricultural pests. A single mango confiscated from an air passenger en route from Nicaragua to Los Angeles was found to contain nine larvae of the destructive West Indian fruit fly. A dozen shamrock seedlings in the bags of two air passengers arriving at Detroit from Scotland had attached soil containing 160 cysts of the golden nematode, a serious pest of potatoes in many European countries. Twenty-eight living larvae of the Oriental fruit fly were removed from four passion fruit found in the baggage of an air passenger at San Francisco from Hawaii. Occasionally pests get through undetected. The Mediterranean fruit fly, an important citrus-fruit pest almost certainly entered the United States through Miami International Airport. The fact that it subsequently spread to 28 counties in the State of Florida and required a 19-month campaign costing 10 million dollars to eradicate it, is a strong argument in favour of measures to check pest movements.

*The sea lamprey*

One of the best documented examples of the biological and commercial impact of an invading species is provided by the entry of the sea lamprey (*Petromyzon marinus*) into the upper Great Lakes of North America (Smith, 1968; Christie, 1974).

Adult lampreys live by attacking fish. They attach themselves to the fish by means of a sucker-like mouth and draw blood and body fluids from their prey (Fig. 6.12). Many of the fish suffering these attacks are killed and only a

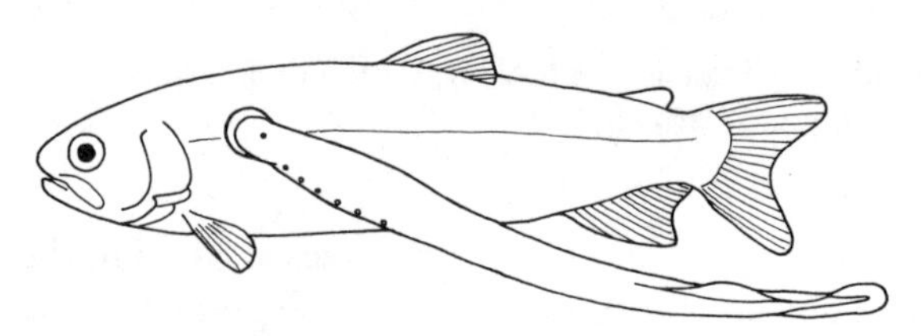

*Fig. 6.12* A lamprey attacking a trout

small proportion survives. The lamprey is primarily an Atlantic coastal species which enters streams in Quebec and the Maritime Provinces to spawn. It has been known to be present in Lake Ontario since the 1830's and may have been there since the glacial retreat. Altertnatively, it may have become established more recently either by passing up the St. Lawrence River or by some less direct route from the Hudson River System. In any event, further penetration into the upper Great Lakes was originally prevented by the Niagara Falls (Fig. 6.13). In 1829 the construction of the Welland Ship Canal provided a route around the barrier of the falls and access into Lake Erie. The lamprey was slow to take advantage of this route probably because of the obstacles presented by the eight large locks of the canal and the strong downstream currents associated with a total fall of

100 m between the Lakes. It was almost a hundred years before the lamprey appeared in Lake Erie, in 1921. Its subsequent spread into the upper Great Lakes was rapid and it reached Lake Huron in 1932, Lake Michigan in 1936 and Lake Superior in 1946. Its impact in Lake Erie was negligible possibly because of the lack of suitable prey and spawning areas. In the other lakes it established itself successfully, spawning in the tributary streams and attacking the local fish populations. The first major fish species to suffer was the commercially important lake trout (*Salvelinus namaycush*) and in each lake as the numbers of lampreys built up, the trout populations collapsed (Fig. 6.13). The collapse was probably aggravated by the intense commercial

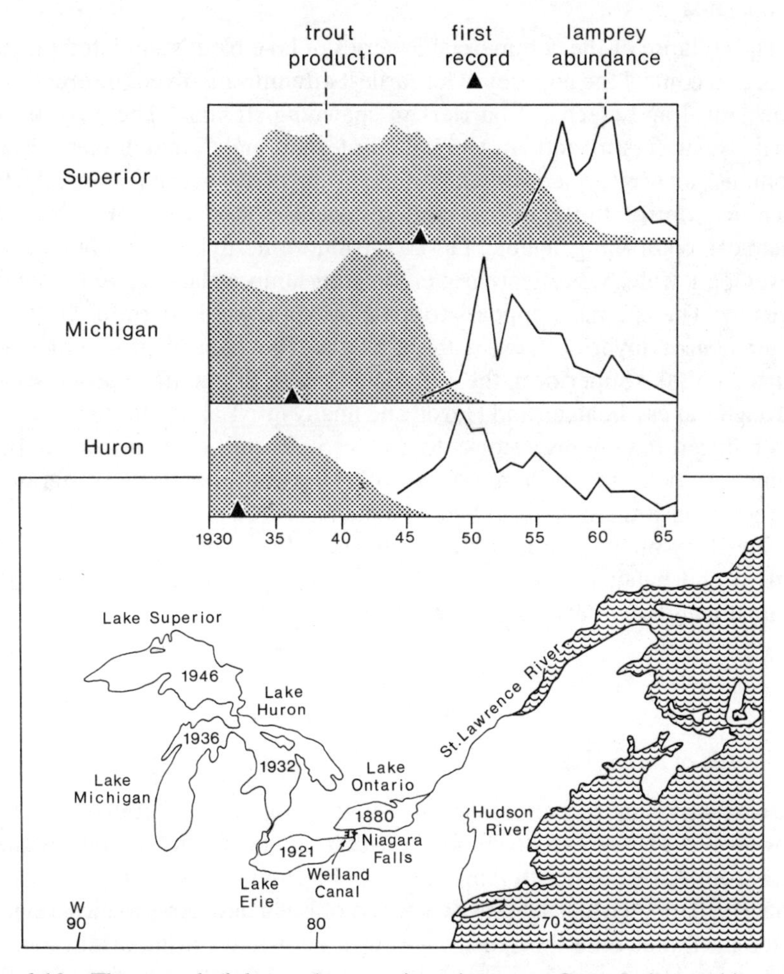


*Fig. 6.13* The spread of the sea lamprey into the upper Great Lakes and its impact on lake trout fisheries

exploitation of the lake trout. These population changes had some further ramifications. In Lake Michigan, as the numbers of lake trout declined both lampreys and commercial fishermen turned their attention to other prey such as the large chubs (*Leucichthys* spp.) which became depleted in their turn. The destruction of the lake trout and the large chubs may have made it easier for another marine invader, the alewife (*Alosa pseudoharengus*) to establish itself. Although this species has some commercial possibilities it suffers periodically from sudden extensive mortalities which in Lakes Huron and Michigan have caused accumulations of dead fish along the shores. This mortality seems to result from the fishes' inability to cope with the high temperatures it encounters when attempting to spawn in the surface waters of the lakes.

The collapse of the commercial fisheries of lake trout stimulated massive efforts to control the lamprey. The earliest attempts involved the erection of mechanical and electrical barriers in spawning streams. These were only partially successful and in 1955 the International Great Lakes Fishery Commission was formed with the principal aim of discovering more effective ways to control the lampreys and restore the lake trout. Eventually a chemical controlling agent, TFM (3-trifluoromethyl-4-nitrophenol) was developed, which was effective in eradicating lamprey larvae from spawning streams. The chemical appears to have no appreciable harmful effects on other fish or invertebrates in the streams. The control programme was started in Lake Superior in the early 1960s and was extended progressively through Lakes Michigan and Huron and finally into Lake Ontario in 1971. It has reduced the lamprey stocks to 10 to 15 per cent of their original peak numbers. Alongside the lamprey control programme has been a programme to re-establish the lake trout using hatchery-reared stocks.

The story of the lamprey illustrates very clearly how the invasion of a settled and balanced community even by a single alien species can have dramatic and far-reaching consequences.

## The spread of disease

### *Transport of animal vectors*

Disease-carrying species make up another important group of animal stowaways. The spread of the malaria-carrying mosquito *Anopheles gambiae* from Africa to South America in 1930 provides a striking example of what can happen when a disease vector colonises a new geographical region.

Up until that time the species was well known as an important malaria vector in Africa but had never been seen in South America. In 1930 however, it appeared on the east coast of Brazil, near the town of Natal. The local airport may have been the point of entry as there was a regular mail-plane service from Dakar on the West African coast (Fig. 6.14). Alternatively, the mosquito could have been carried across the Atlantic on a mail ship. In any event, a serious epidemic of malaria followed, starting near Natal and spreading northwards into the previously malaria-free valleys of Assi, Mossoro, and Jaguaribe (Soper and Wilson, 1943). In all, the epidemic affected hundreds of thousands of people and caused 14 000 deaths in an area of 30 000 km$^2$. Although malaria carried by other mosquitoes had been known for many centuries in South America, the arrival of *Anopheles gambiae* had a particularly devastating effect. Its efficiency in transmitting the disease is probably related to the habit of breeding outside the forest, using open sunlit pools in clearings close to settlements.

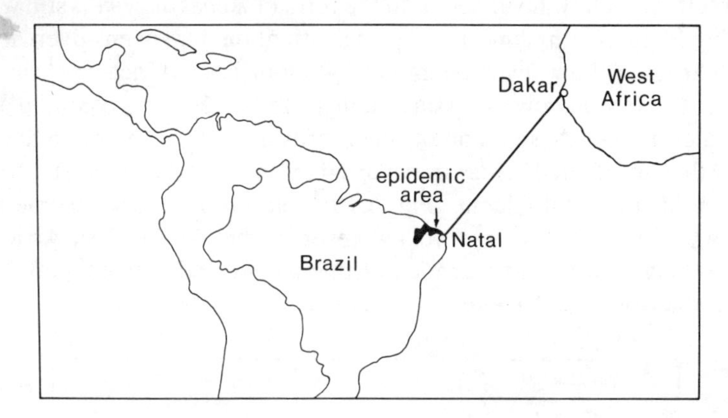


*Fig. 6.14* The invasion of Brazil by *Anopheles gambiae* from West Africa

Fortunately an energetic campaign mounted by the Brazialian government and the Rockefeller Foundation, using chemicals against adults and larvae, succeeded in eradicating the mosquito in under two years, albeit at the cost of some two million dollars. The success of these measures, one of the few examples of successful eradication of a disease-vector, was probably related in part to the small number of original colonisers. This would have limited the genetic variability in the resultant population and restricted its capacity to adjust to new conditions or to become resistant to insecticides. Following this episode, careful searches and spraying measures were instituted in Brazil for incoming ships and aircraft and it is interesting to note that in a survey during 1941 and 1942 seven more *Anopheles gambiae* adults were recovered from aircraft arriving from Africa.

Various surveys have shown that the carriage of species of potential public health significance is by no means a rare occurrence. Searches of aircraft arriving at Nairobi airport in Kenya showed that in two months (March and

April 1968) the 27 aircraft examined yielded no less than 150 adult mosquitoes. Similar searches of 65 aircraft, between November 1968 and July 1969 produced 340. In both cases a large proportion of the species captured were potential disease-carriers (Highton & Van Someren, 1970). Similarly, when goods trains travelling into North America from Mexico were examined at Brownsville, Texas, over 3000 mosquitoes were taken from 173 trains over a two-year period (1958–1960). These included a number of malaria vectors.

Certain combinations of routes and vectors are of special significance. For example, in North America there is a continual risk of malaria being reintroduced from Mexico and of yellow fever being reintroduced from South America or Africa. Yellow fever once occurred as far north as Boston and a suitable yellow fever vector in the form of *Aedes aegypti* is still widely distributed in North America. Special attention has been given to the possibility of yellow fever being carried from East Africa to Asia. The disease is so far unknown in Asia although *Aedes aegypti* is common and is implicated in the transmission of other virus diseases such as dengue. It is feared that an infected *Aedes* transported by an aircraft from East Africa to Asia could start an epidemic (Fig. 6.15). The problem has become more pressing with recent severe yellow fever outbreaks in East Africa. In 1960–62 a yellow fever epidemic in Ethiopia affected 200 000 people and caused the deaths of 30 000.

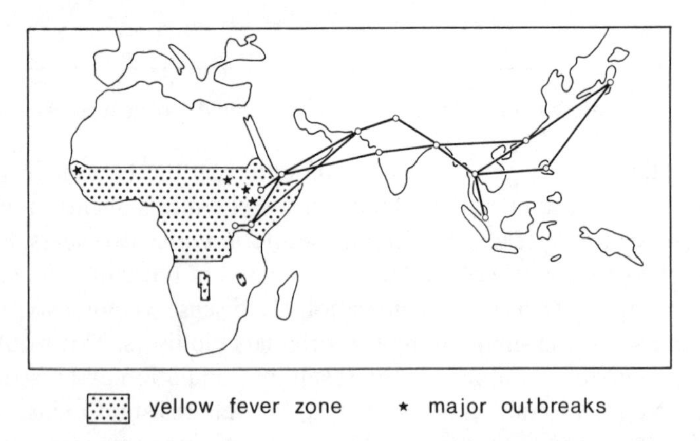


*Fig. 6.15* Air routes linking the yellow fever zone of Africa with disease-free areas in Asia

One might question why yellow fever has not reached Asia already since the dhows which have plied between East Africa and Asia for centuries must surely have carried both yellow fever sufferers and drinking water vessels containing *Aedes* larvae. The failure of yellow fever to establish itself by this route possibly argues in favour of some degree of immunity amongst the human populations of Asia. Such immunity could have arisen as a result of infections by an antigenically similar virus, such as the one causing dengue fever. However, even if such a mechanism exists it is unlikely to be entirely effective and in no way removes the need for procedures to prevent the carriage of live mosquitoes in aircraft.

New transport routes, such as the new roads in the Amazon basin are likely to produce additional problems. There is, for example, an important focus of Chagas disease around the east coast terminus of the Trans-Amazon Highway (Fig. 6.10) and the vectors for the disease, various species of reduviid bugs, travel readily on motor vehicles and may be expected to carry the infection westward into the interior. Similarly the western Brazilian terminus of the Perimetral Norte route is close to one of the few Brazilian sites for onchocerciasis. The disease vector, a blackfly, is known to be capable of travelling long distances on motor vehicles and this is likely to facilitate the spread of the disease.

Technological developments can also introduce new hazards. Because of their involvment in plague transmission there are well-tried procedures for controlling rats in ports. However, the excellent record of plague and rat control in recent years may be threatened by the use of 'containerised' cargo systems. The containers tend to be packed and unpacked at factories, granaries and warehouses and unless special provisions are made, this system will by-pass the established procedures for rat control at ports.

### *Control methods*

There are basically two strategies for checking the spread of disease-carriers. One is to try to prevent them from approaching or leaving vehicles at centres involved in international transport. The other is to attack them in transit. The principles are much the same whether the vehicle is an aircraft, a ship, a train or a lorry and whether the centre is an airport, a seaport or a road and rail terminus. The appropriate procedures have been fully described in a useful manual on 'Vector Control in International Health' (World Health Organisation, 1972b).

The control strategy for an international airport in the tropics could include the following measures. An attempt should be made to clear all mosquitoes from the airport by spraying with pesticides, removing such objects as old tins which might collect water and provide breeding places for *Aedes aegypti* and possibly by introducing mosquito-eating fish into cisterns and ornamental pools. Rat-control measures should include laying down of poison baits, rat-proofing of buildings and the effective disposal of refuse. These operations would be extended where possible in a 'cordon sanitaire' of at least 400 m around the perimeter and their effectiveness continually monitored using devices such as light traps and egg traps for mosquitoes, and spring traps for rats.

However effective these measures might seem it must always be assumed that rats and mosquitoes could still have entered the aircraft. Whereas poison baits are suitable for rat control on ships, traps are used on aircraft, otherwise there is the danger of the poisoned animal crawling away and becoming lodged in some important part of the mechanism. For mosquito control the method currently used, although not applied in all the situations where it would be appropriate, is for insecticide sprays to be discharged into the cabins of aircraft as they are taxiing for take-off (Sullivan *et al.*, 1972). The usual practice is for the cabin crew to walk down the aisles discharging the spray from aerosol cans. Even high capacity aircraft such as the Boeing 747 'jumbo jet' can be treated by four stewardesses in less than one minute. It is claimed this method kills between 99 and 100 per cent of the mosquitoes. The insecticide used is the synthetic pyrethroid *resmethrin* mixed with a suitable propellant. Some initial difficulties have been experienced because the residues from the insecticide under the action of sunlight developed a musty odour after a few days. This problem is overcome by using a more highly refined product.

### *Diseases carried by imported animals*

It is of course not only stowaway species such as mosquitoes and rats which can transmit diseases to man. There can be health risks associated with intentional introductions. Parrots and parakeets imported for the pet trade can transmit psittacosis to man and it is claimed that the particularly virulent strains are likely to become overt in birds exposed to the stresses of importation. Monkeys imported for medical research can act as carriers of Marburg disease, a recently recognised virus infection which in 1967 was responsible for 31 cases and 7 deaths amongst laboratory workers and their contacts in Germany and Yugoslavia. Of particular concern in Britain is the possibility that rabies will be reintroduced by a pet-owner evading the animal quarantine regulations. This would endanger not only man himself but also his pets, his domestic stock and many wild species.

### *Man as a disease carrier*

Any account of transport and disease would be incomplete without a consideration of man himself as a disease carrier. One of the first measures aimed at preventing the international spread of disease was the detention of ships, cargoes, crews and travellers for 40 days in Mediterranean ports during the 14th century plague epidemic; a custom which gave us the word 'quarantine' from the Italian word meaning 'forty'. Since that time the causes and mode of transmissions of the major diseases have been understood and immunisation techniques have been developed for many of them. Strategies have also been formulated to check the world-wide spread of disease and have been embodied in the International Health Regulations (World Health Organisation, 1974a).

Unfortunately these provisions are far from foolproof particularly in an age of fast air travel (Dorolle, 1968). Newly recognised diseases still take us by surprise as was amply demonstrated by the recent carriage by air travellers of the Lassa fever virus from Africa to the United States and Britain (Monath, 1975). Even with well-known diseases the system sometimes breaks down; smallpox is a case in point. Vaccination for the disease is entirely reliable, yet between 1960 and 1970 smallpox was introduced into Europe on 28 different occasions with 552 cases reported from 12 different countries. Almost all the carriers had travelled by air from endemic areas in Africa, Asia and South America. The conclusion to be drawn here is that the possession of a vaccination certificate is not necessarily a guarantee that the holder has actually been vaccinated.

There are even more difficulties in the case of cholera. The immunity produced by immunisation is transitory and persists for only three to four months (compared with three years for smallpox and nine to ten years for yellow fever). The causative organism can be carried without symptoms for a year in the case of classical cholera (caused by *Vibrio comma*) and three years in El Tor cholera. Finally many potential cholera carriers travel by smuggling and other uncontrolled routes and thereby escape the jurisdiction of international health procedures. This loophole proved particularly significant in the 1961–1973 El Tor cholera epidemic (Fig. 6.16). The disease started in the Celebes and the transport of infected persons on small coastal vessels was certainly a significant factor in its initial dispersal. By 1973 it had spread widely in Europe and Africa, invading in the process countries which had been free from cholera for many decades.

The spread of influenza represents a different kind of problem because the virus can appear in a wide range of different forms, each of which requires a different antiserum to combat it (Stuart-Harris, 1965). Consequently it is rarely feasible to immunise travellers in advance. Control has had to be organised on a different basis and this has involved the establishment of a world-wide surveillance programme co-ordinated by the World Influenza Centre in London. This works by collecting and disseminating information on the distribution of the disease and on the antigenic structure of the current epidemic virus strain. The programme was given an opportunity of showing its efficiency in the 'Asian flu' epidemic of 1968. This first attracted

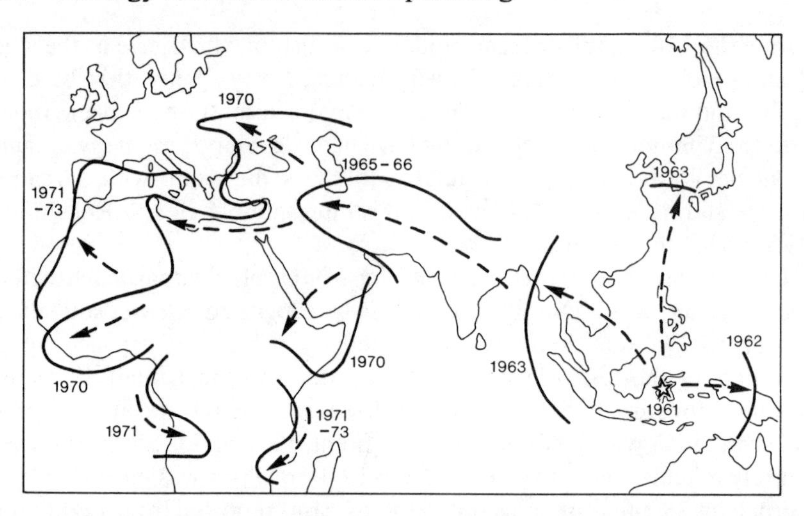


*Fig. 6.16* The spread of El Tor cholera in the 1961–1973 epidemic (based on a World Health Organisation map)

attention in early July when about 500 000 cases occurred in Hong Kong. Even before the epidemic had reached its peak the strain had been isolated by the local influenza centre and flown to the world centre, where within two weeks it had been characterised, freeze-dried, and made available to vaccine-production laboratories.

# Case studies

*Planners need ecological information not only about the impacts of individual developments, but also about the likely interactions between all the enterprises represented in any particular region. Although in detail the number of possible combinations of land uses is very large, it is possible to recognise a relatively small number of basic themes or planning strategies. At one end of the scale are situations dominated by urban and industrial development, at the other end are areas such as National Parks where priority is being given to landscape protection and conservation. Each planning strategy requires its own kind of ecological input and four different situations are described in the chapters which follow. The first three relate to adjacent but contrasting regions of South Wales (Fig. 7.1), the fourth to a situation in a developing country.*

# 7 A coastal development area

During the last 40 years the most striking industrial development along the South Wales coast has been the growth of modern iron and steel, and petrochemical industries (Fig. 7.2, *1–6*). During the same period new urban growth centres and light-industrial estates (*7–13*) have also been established to redeploy the labour force released by the decline of the coal industry. Ports such as Swansea, Barry, Cardiff and Newport, originally established

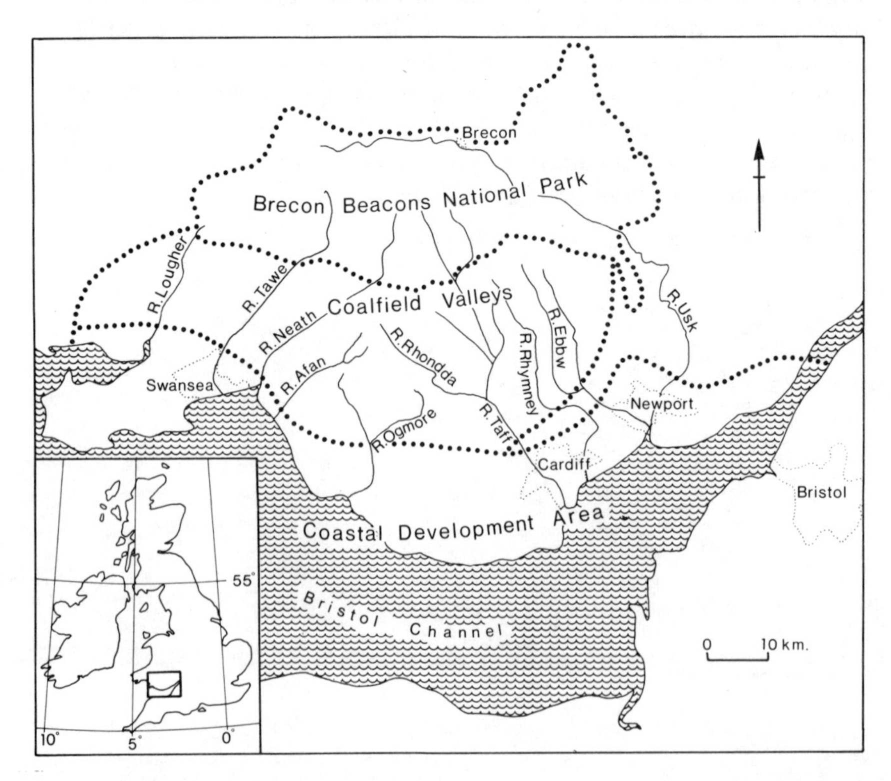


*Fig. 7.1* The South Wales study areas

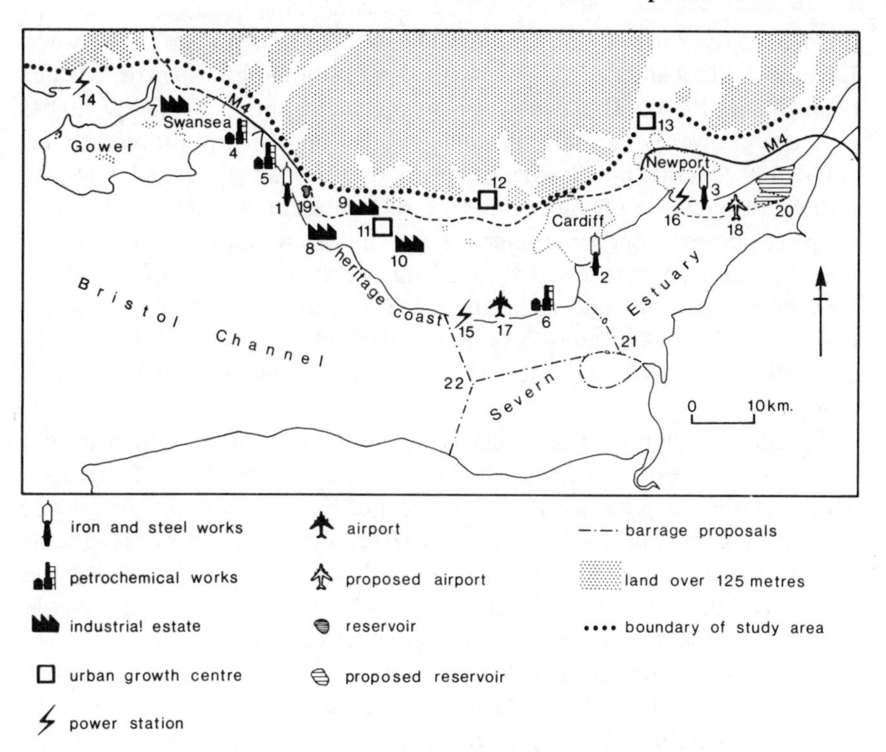


*Fig. 7.2* Existing and proposed developments along the South Wales coast

*Iron and steel works* — *1.* Margam (Port Talbot), *2.* Cardiff, *3.* Llanwern
*Petrochemical works* — *4.* Llandarcy, *5.* Baglan Bay, *6.* Barry;
*Industrial estates* — *7.* Fforestfach, *8* Kenfig, *9.* Bridgend, *10.* Waterton
*Urban growth centres* — *11.* Bridgend, *12.* Llantrisant, *13.* Cwmbran,
*Power stations* — *14.* Burry Port, *15.* Aberthaw, *16.* Uskmouth,
*Airports* — *17.* Rhoose, *18.* Severnside (proposed),
*Reservoirs* — *19.* Eglwys Nunydd, *20.* Severnside (proposed),
*Barrage proposals* — *21.* Lavernock point to Brean Down, *22.* Aberthaw to Watchet

for coal-shipping, have now been reorganised to serve the new industries and a deep water harbour for iron-ore carriers has been built at Port Talbot.

The estuary itself offers many opportunities for development. Its tidal range, the second largest in the world makes it a very suitable location for a barrage incorporating a tidal power station. The idea of a Severn barrage is not new; it was first mooted in 1933 when a proposal was made to construct a barrier across the estuary near the position of the present Severn bridge. A much more ambitious scheme put forward in 1945 would have involved a construction more than 20 km long across the lower part of the estuary from Aberthaw to Watchet (22). In this scheme the lower estuary was to be divided into two basins. The most recent variant proposes a barrage between Lavernock Point and Brean Down with a small adjoining basin in deep water (21) (Shaw, 1974). The two-basin arrangement allows flexibility in the pattern of power generation. A single straight dam incorporating turbines could only generate power when strong tides were moving up and down the estuary through the turbine tunnels. Using an additional lower basin and off-peak power from conventional power stations, water could be pumped into the basin, then released through the turbines to compensate for the 'generating gaps' in the normal tidal cycle. The small installation at La Rance in Brittany has already demonstrated the feasibility of tidal power generation. A Severn barrage would produce at least ten times as much power.

Another development prospect is to create new building land by the reclamation of estuarine mud flats. One ambitious scheme of this type involves converting the large mud flat east of Newport into an airport and reservoir complex (*18*, *20*) (Hooker, 1970). Finally the low-lying agricultural land between Cardiff and Chepstow has been recognised as having great possibilities for development and has been considered as a possible location for a Maritime Industrial Development Area (MIDA).

Although proposals for development may be in the national interest they usually give rise to conflicts. Any kind of construction inland is likely to involve encroachment on 'greenfield' sites valued for other purposes: agriculture, recreation, wildlife conservation or, more subtly, landscape quality. Equally, estuarine barrages or reclamation schemes are liable to interfere with conservation and fishery interests. The discharge of urban and industrial wastes raises the possibility of hazards to health and to crops.

## Encroachment on rural land

Developments along the coast have resulted in a progressive loss of rural land starting with the major expansions of Cardiff, Newport and Swansea in the 19th century and extending up to the present time with the construction of the M4 motorway.

The development of the modern iron and steel industry, first at Margam in the early 1940's and subsequently at Llanwern in 1962, destroyed large areas of low-lying farmland along the coastal strip. This was land largely reclaimed from the sea in mediaeval times and originally consisted of a series of 'moors', with Baglan and Margam moors near Port Talbot, Wentloog moors between Cardiff and Newport and Caldicot moors east of Newport. The Margam steelworks and the Baglan Bay petrochemical complex have virtually destroyed the western moors, but apart from the intrusion of the Llanwern steelworks, the eastern moors have remained virtually intact.

This land is of considerable importance in its rural state. It lies on fertile alluvial clays and is well-suited to dairy farming and stock rearing. Its drainage is based on a complex series of ditches or 'reens' which also serve as field boundaries and watering places for stock. It is the reens with their associated reed-beds and lines of pollarded willows which give the area its particular landscape character. The channels also have considerable conservation interest because they support a number of water plants which have become restricted elsewhere in Britain. These include arrowhead (*Sagittaria sagittifolia*), flowering rush (*Butomus umbellatus*) and frogbit (*Hydrocharis morsus-ranae*). The pastures and wet areas on the moors are a traditional feeding ground for wildfowl. Before the Margam moors were lost to the steelworks they were visited by 1500–2500 overwintering white-fronted geese (*Anser albifrons*), a large proportion of the European race of this species (Heathcote *et al.*, 1967). The remaining moors are still important as roosting areas for waders. For all these reasons, further industrial development on the moors is likely to meet with opposition from many quarters.

*New wildlife habitats*

When cataloguing conservation losses it is easy to lose sight of the fact that industrial development can sometimes involve compensatory gains. A notable example is the Eglwys Nunydd Reservoir constructed for the Margam steelworks (Fig. 7.2, *19*). This has developed into an eutrophic lake (p. 84) and provides an excellent habitat for a wide range of overwintering ducks (Table 7.1). In this way it helps to make up for the loss of waterfowl refugees on Margam moors. The coal-burning power stations at Aberthaw and Uskmouth have also produced a conservation bonus in the form of fuel-ash lagoons. These provide high-tide roosts for wading birds and the ones at

*Table 7.1* Average winter populations of ducks on Eglwys Nunydd Reservoir 1963–1973

| | *Sept.* | *Oct.* | *Nov.* | *Dec.* | *Jan.* | *Feb.* | *Mar.* |
|---|---|---|---|---|---|---|---|
| Wigeon | – | 2 | 4 | 4 | 4 | 5 | – |
| Teal | 1 | 2 | 14 | 25 | 39 | 37 | 18 |
| Mallard | 3 | 4 | 14 | 3 | 12 | 11 | 5 |
| Pochard | 12 | 59 | 224 | 241 | 256 | 185 | 79 |
| Tufted duck | 13 | 42 | 120 | 139 | 140 | 108 | 66 |
| Goldeneye | – | – | 3 | 6 | 5 | 7 | 13 |
| Shoveler | – | – | – | – | 6 | 16 | 8 |
| Other species | – | – | 4 | 4 | 1 | 5 | – |

Uskmouth represent the most important roost on the north side of the estuary.

## Estuarine developments and conservation

Most estuaries have a significance beyond their development potential. In temperate regions many of them represent important habitats for overwintering geese, swans, ducks and wading birds and also serve as thoroughfares for migratory salmon.

In the Severn estuary any proposal which substantially reduces the area of mud flats exposed at low tide is likely to have an adverse effect on the birds. On the north side dunlin and knot are the dominant species, with smaller numbers of turnstone, ringed and grey plover, curlew, redshank and shelduck (Figs. 7.3, 7.4). Bridgwater Bay is particularly significant as a moulting area for shelduck. Further west in areas such as the Burry Inlet the oystercatcher, much discussed in relation to the cockle fishery, is the important species. The Severnside reclamation and airport scheme would largely eliminate the major feeding area on the north shore (Fig. 7.4: *a*). The barrage would also reduce the total area of mud exposed at low tide by decreasing the tidal range (probably from 15 to 6 m maximum).

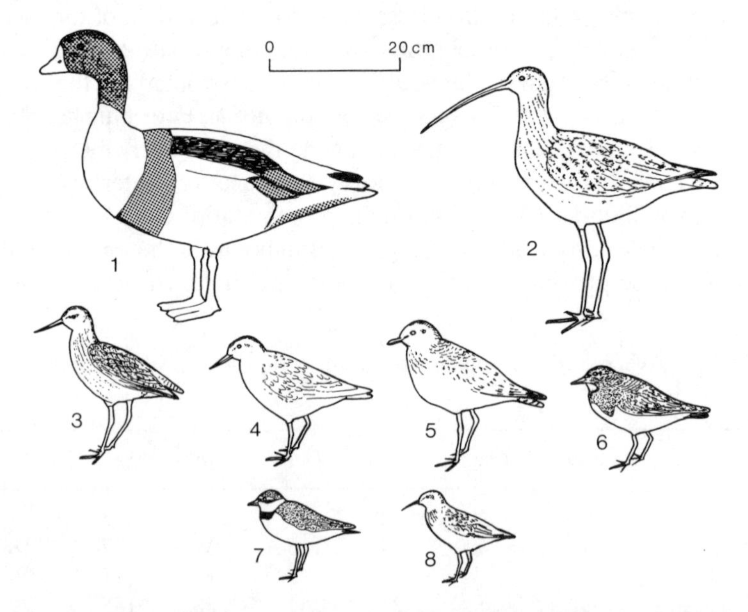


*Fig. 7.3* Common overwintering species in the Severn estuary: *1.* Shelduck (*Tadorna tadorna*); *2.* Curlew (*Numenius arquata*); *3.* Redshank (*Tringa totanus*); *4* Knot (*Calidris canutus*); *5* Grey plover (*Pluvialis squatarola*); *6. Turnstone (Arenaria interpres); 7.* Ringed plover (*Charadrius hiaticula*); *8.* Dunlin (*Calidris alpina*)

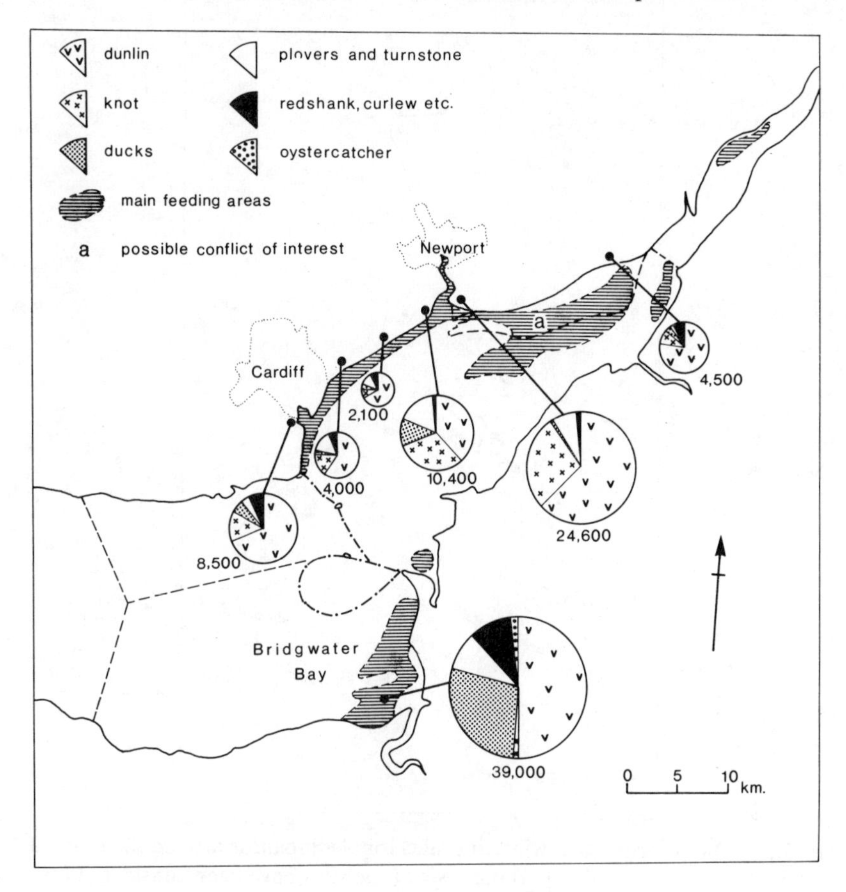


*Fig. 7.4* The distribution of wading birds and ducks at roost sites along the northern shore of the Severn estuary and at Bridgwater Bay (counts provided by A. J. Prater, British Trust for Ornithology)

It is difficult to predict the overall effect on the birds. Either development would certainly bring about a significant reduction in numbers. It is unlikely however that any of the major species would disappear altogether from the mud flats. The species seem to be ecologically separated, not so much according to geographical localities along the shore, but rather on the basis of the mud and water depths appropriate to their individual beak and leg lengths (Edington *et al.*, 1973). None of these feeding options would be lost by a reduction in total area of the mud flats.

From the results of the 'Birds of Estuaries Enquiry' (Prater, 1974), Britain's estuaries can be ranked in order of importance as overwintering sites for waders. (Fig. 7.5). On this basis the Severn estuary is currently ranked eighth. Even if its wader populations were halved, it would still

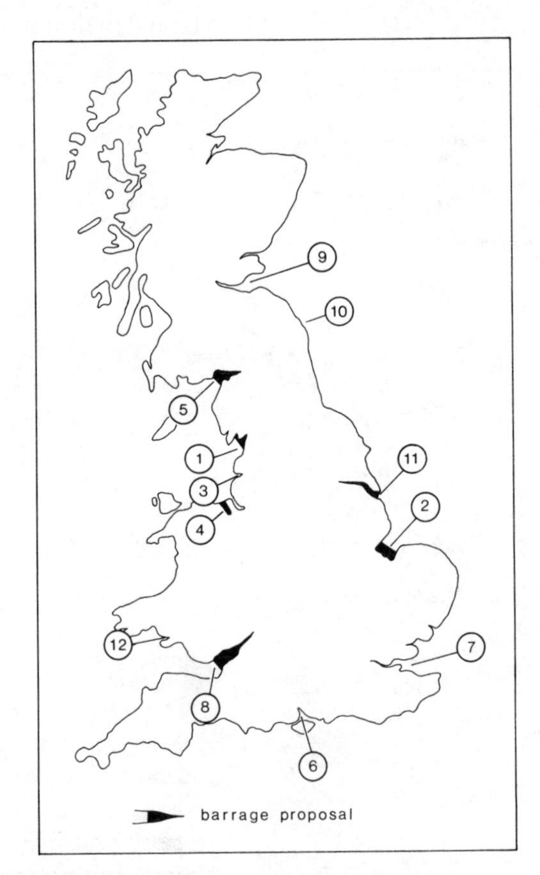


*Fig. 7.5* The principal overwintering sites for wading birds around the coast of Britain, listed in order of importance. Six of the sites have been considered for barrage schemes of various kinds (wader counts from Prater, 1974).

| *Estuary site* | *Peak wader count 1972–1973 (thousands)* | *Nature of barrage proposal* |
|---|---|---|
| *1.* Morecambe Bay | 233 | Freshwater reservoir |
| *2.* Wash | 165 | Freshwater reservoir |
| *3.* Ribble | 159 | |
| *4.* Dee | 154 | Freshwater reservoir/ road crossing |
| *5.* Solway | 144 | Freshwater reservoir/ tidal power station |
| *6.* Hants/Sussex Harbour | 100 | |
| *7.* Thames | 84 | |
| *8.* Severn | 76 | Tidal power station/ navigation aid |
| *9.* Firth of Forth | 63 | |
| *10.* Lindisfarne | 39 | |
| *11.* Humber | 38 | Navigation aid |
| *12.* Burry Inlet | 35 | |

support nearly 40 000 birds and be rated twelfth in the list of 32 major sites, each supporting more than 15 000 birds. The implementation of development schemes in other estuaries could of course change the ranking order and aggravate the conservation problem overall. However it is unlikely on financial grounds that more than one of these major schemes will be undertaken in the near future.

## Estuarine developments and fisheries

Conflicts between industrial development and salmon fisheries are not a new problem in the rivers draining into the Severn estuary. On the Welsh side, the salmon fisheries in the Rivers Afan, Taff, Rhymney and Ebbw (Fig. 7.1) were all destroyed by a combination of industrial pollution and physical obstructions. More recently a conflict arose on one of the surviving salmon rivers, the River Usk, as the result of the extraction of power station cooling water from its tidal reaches (Fig. 7.6*a*). When the Uskmouth power station was enlarged in 1962 by the addition of a second generating unit (the B

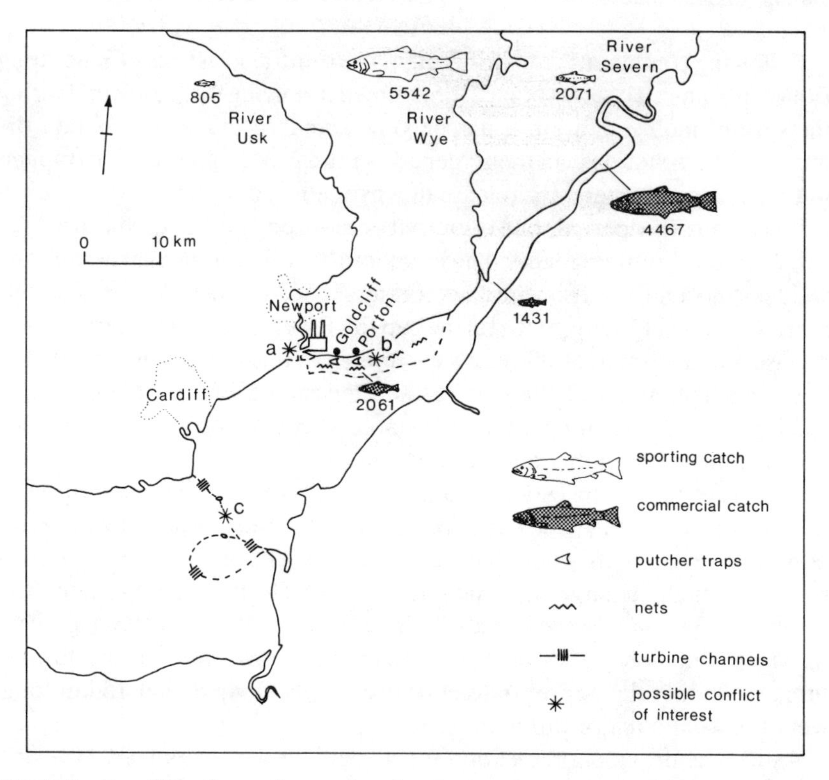


*Fig. 7.6* Possible effects of estuarine developments on salmon fisheries

station) it became evident that the new water extraction arrangements were interfering with the downstream migration of salmon smolts and in the first year of operation approximately 5000 were killed. The problem was traced to the pumps on the new B station which had been located in front of the debris screens rather than behind (Fig. 7.7). Fish drawn into the pumps were being damaged by blows from the rotor blades and by the pressure extremes always associated with pumping machinery.

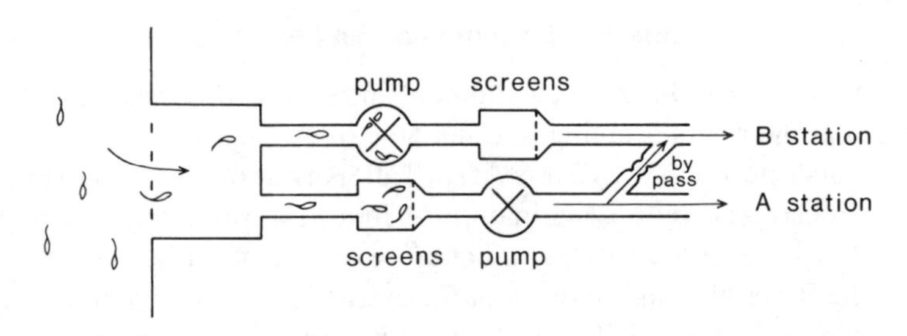


*Fig. 7.7* The arrangement of cooling-water intakes at the Uskmouth Power Station (diagrammatic)

Following complaints from fishermen, the Central Electricity Generating Board introduced a series of measures aimed at solving the problem. During the spring smolt migration it has become standard practice to obtain the water for both stations via the screened A intake and to rescue fish trapped in the screen chambers, for release downstream. To reduce water requirements during this period, power output is reduced and the time is used for carrying out routine maintenance operations. To compensate for any residual mortality of fish, the Generating Board has established a smolt-rearing station at Llanfrynach further up the River Usk which guarantees to release a minimum of 5000 smolts each year. In this way a compromise has been reached, albeit at the cost to the Generating Board of operating a rearing station, mounting an annual smolt rescue operation, and reducing power output during the migration period.

A barrage across the estuary would present a very considerable obstacle to fish movement (Fig. 7.6*c*). From evidence obtained at hydroelectric dams built on rivers, smolts would have no particular difficulty in moving downstream through the large slow-moving turbines. The prospects for adult fish returning upstream are more difficult to predict. Whether the fish were provided with fish passes or were expected to move through the turbine tunnels, finding the appropriate entrance points along a wall 15 km long, would present a major problem for them.

As well as providing recreation for anglers on the rivers Usk, Wye and Severn, the salmon also support a commercial fishery which, on the north side of the estuary, can yield 8000 fish per annum (Fig. 7.6). The fishery makes use of basket traps known locally as 'putchers' and an assortment of stop and drift nets. At Goldcliff, near Newport, the arrangements for trapping salmon may involve as many as 2000 baskets set into fences (Fig. 7.8). This particular enterprise would certainly be displaced by the Severn-

*Fig. 7.8* The basket traps for salmon at Goldcliff. Surprisingly these traps work best when facing upstream.

side reclamation and airport scheme (Fig. 7.6*b*). Whether it could be successfully relocated at the outer edge of the reclaimed area is a matter for conjecture in view of our very limited knowledge of fish movements and the likelihood that current patterns would be substantially altered by any reclamation project.

## The effects of industrial wastes

Although the most important agricultural area, the Vale of Glamorgan is out of range of the effluents from the steel and petrochemical industries, concern has been expressed about more local effects on farms and forest plantations. However, very often it is impossible to separate the effect of industrial pollutants from other influences. Downwind from the Margam steelworks there are some groups of dead and dying trees which have become much favoured by photographers to illustrate the effects of atmospheric pollution. In fact many of these trees were destroyed by a local fire and they are all in a position which renders them vulnerable to salt-spray damage. Lichens are now recognised as sensitive indicators of pollution by sulphur dioxide (Hawksworth & Rose, 1970) and various lichen studies have shown that the diversity of species decreases in the vicinity of the steelworks (Pyatt, 1970). However, the implications for the adjacent forest plantations are again not clear. Whilst many tree species show poor growth downwind from the steelworks, they also show similar poor growth elsewhere in South Wales on sites where pollution is unlikely to be a contributory cause.

Virtually all the water-borne effluents from the coastal industries are discharged directly into the tidal waters of the estuary. Swansea Bay, between Mumbles and Sker Point, receives over 210 000 $m^3$ of industrial waste daily and the Severn estuary, between Lavernock Point and the Severn Bridge, 42 000 $m^3$ (Department of the Environment, 1974*b*). The main discharges on the northern side are from the iron and steel works at Port Talbot, Cardiff and Newport, the petrochemical works at Port Talbot and Barry, and from a paper mill at Sudbrook. There is a similarly varied range of discharges on the south side. In 1972, a number of studies on the water in the estuary and the animals and plants along its shores revealed higher concentrations of heavy metals than occur in more open sea areas (Abdullah *et al.*, 1972; Butterworth *et al.*, 1972; Nickless *et al.*, 1972; Preston *et al.*, 1972). For example, dog-whelks ((*Nucella lapillus*) along the northern shore have been found to contain as much as 4200 p.p.m. zinc, 725 p.p.m. cadmium and 38 p.p.m. lead. This can be compared with values of 850 p.p.m. zinc, 155 p.p.m. cadmium and 2 p.p.m. lead further west at Dale in South Pembrokeshire. The level of cadmium in one sample of the alga *Fucus vesiculosus* from Barry was the highest found in a survey of British coastal waters. There is little evidence however that any animal or plant species is declining as a result of pollution. Samples of fish taken from the intake screens at the Oldbury nuclear power station show a diversity of species which compares favourably with a faunal list for the area compiled in 1940 (Hardisty *et al.*, 1974). Likewise the records for salmon catches on the Rivers Usk, Wye and Severn, give no indication of any downward trend. Industrial pollution could be implicated in the recent decline of the cockle industry in the Burry Inlet, but the situation is undoubtedly complicated because in addition to the hypothesis involving increased predation by oystercatchers, changes in sediment distribution and invasion of the cockle beds by *Spartina* grass could also be important factors.

It is well known that man can be adversely affected by industrial pollutants, particularly by materials released into the atmosphere and by toxic substances accumulating in the organisms he uses for food. Both these possibilities have been under scrutiny in South Wales. The accumulation of heavy metals in shellfish is one of the potential hazards but the analyses carried out to date show that the levels are still within safe limits. Atmospheric pollutants are undoubtedly a source of annoyance and inconvenience in the region; red dust from the Margam steelworks, greasy soot from the Port Tennant carbon black factory, and unpleasant odours from the various petrochemical works have all attracted adverse reactions from householders. So far, however, no situation has been identified which represents a proven hazard to health.

## The effects of urban wastes

Many of the problems generated by urban development are ones associated with domestic wastes. Air pollution from coal-burning on domestic fires has never been a serious problem in South Wales, because although coal was extensively used it was mainly of a low-volatile type which produced little smoke. The region has not however escaped those difficulties so frequently associated with the disposal of sewage and solid domestic wastes.

The doubts now being raised about the desirability of discharging partially treated sewage into recreational waters have particular relevance to the South Wales coast, which boasts a number of major holiday resorts but also receives discharges of crude sewage amounting to 105 000 $m^3$ daily (1974). In several places sewage issues directly on to beaches from cracked and broken pipes and two recent studies (1974, 1976) have shown that the bathing waters at 26 out of 45 beaches had a total coliform count exceeding the proposed E.E.C. standard of 10 000/100 ml. Indeed three beaches had counts which exceeded even the 800 000/100 ml standard advocated by the U.S. Environmental Protection Agency, by far the most permissive of all the values proposed (p. 45).

Although contamination of coastal bathing waters has received most attention it is becoming increasingly apparent that inland recreational waters also need to be considered. Many rivers and streams in South Wales used informally by children are contaminated by discharges from overloaded sewage systems. Most of the offending discharges originate from sewers which combine the function of both foul-drainage and surface-water disposal. Many of these have become overloaded as a result of urban growth and increased domestic use of water. This results in regular discharges of diluted but untreated sewage into surface watercourses, particularly following rain when there is an influx of storm water.

The city of Cardiff, although rightly renowned for its progressive approach to amenity planning, suffers from these problems in the same way as other urban centres in the region. Figure 7.9 gives results of a bacteriological examination of streams and rivers associated with parks and playing fields in the city. The stream in Heath Park, for example, gave faecal coliform counts up to 256 000/100 ml and in the other park areas, regular counts in excess of 20 000/100 ml were quite usual. It will be recalled (p. 44) that faecal coliform counts are more discriminating than total coliform counts in detecting faecal contamination, and subsidiary tests using the faecal coliform/faecal streptococci ratio confirmed that the contamination in the parks was predominantly human in origin. Every site tested greatly exceeded the proposed E.E.C. standard for faecal coliforms in freshwater (2000/100 ml).

Unfortunately, interpreting the public health significance of such departures from recommended standards will remain a problem until a comprehensive medical study is made and published of the circumstances likely to lead to bathers and other water-users contracting infections.

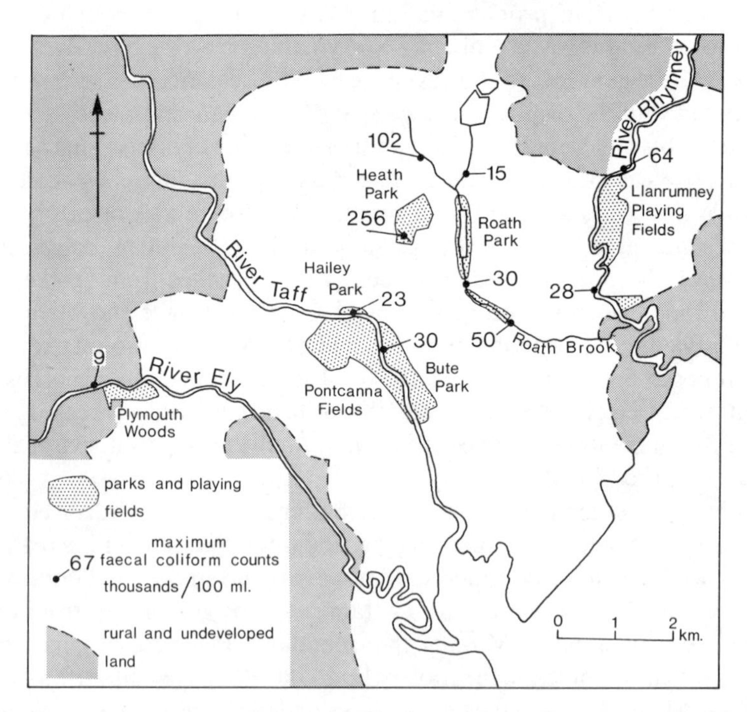


*Fig. 7.9* A survey of faecal contamination in water supplies associated with Cardiff City Parks (from a survey by W. Bray, University College, Cardiff)

## *The eutrophication of reservoirs*

Apart from health implications, incomplete sewage treatment can lead to management problems in reservoirs. When water enriched by the breakdown products of sewage, particularly phosphates and nitrates, is impounded in a reservoir, it is likely to support dense growths of algae which are difficult to remove from the water.

The South Wales coast provides a classic example of eutrophication at the Eglwys Nunydd reservoir of the Margam steelworks. The reservoir was constructed in 1962 to supplement local river supplies which were inadequate in dry summer periods. However, much of the extra water for the reservoir is abstracted from the nearby River Kenfig at a point about three km downstream from the discharge of an overloaded sewage works. Not surprisingly the nutrient-rich water gives rise to algal blooms on the reservoir in summer. The important bloom-forming species in this case is the blue-green alga *Anacystis aeruginosa*. The situation is further complicated by the fact that the prevailing winds cause an accumulation of algae at the extraction point. This combination of circumstances can make the reservoir virtually unusable in the summer months, a situation which could threaten steel production in the absence of alternative water sources.

Problems of this type are not uncommon in lowland industrial areas where it is necessary to supplement good water supplies from upland catchments with water drawn from the enriched lower courses of rivers. In these circumstances the possibility of troublesome algal growths needs to be recognised at the design stage to enable suitable remedial measures to be incorporated. These may include: providing high capacity filtration facilities; arranging for the removal of nutrients from the water before it enters the reservoir; or building a series of separate basins served by different water supplies to reduce the chance of blooms occurring on them simultaneously.

### *Refuse tips, gulls and airports*

Solid domestic refuse is the other major form of urban waste and gulls attracted to urban refuse tips have become a serious pest at many airports in temperate regions (p. 113). As elsewhere the gull populations along the South Wales coast have increased in numbers over the last 20–30 years probably as a result of their increasing use of refuse tips. The principal species involved are the herring gull (*Larus argentatus*) and the lesser black-backed gull (*Larus fuscus*). Most of the resident birds breed on the mid-channel islands of Steep Holm and Flat Holm. On Flat Holm, which was colonised by gulls in 1954, the number of breeding pairs had reached 1600 by 1966 and 6000 by 1974.

At all seasons the gulls make regular daily journeys between roosting sites and the urban areas where they obtain food from refuse tips. Figure 7.10 shows some typical movement patterns for gulls in winter. Gulls using the refuse tips in Cardiff and in the Ely, Taff and Rhymney Valleys move out

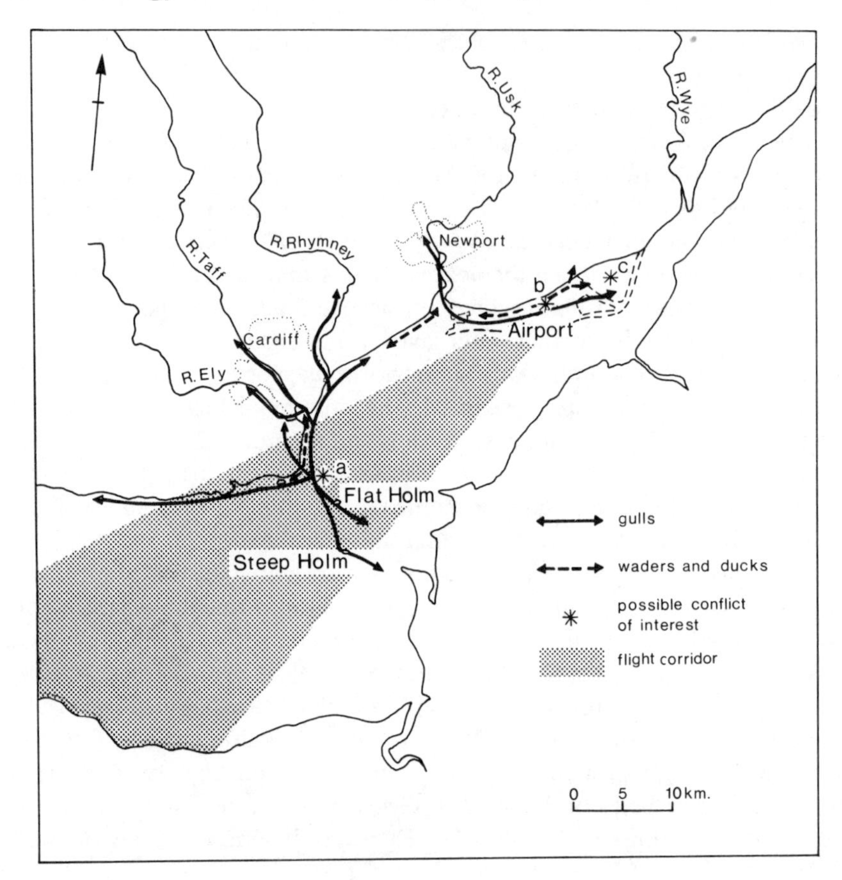


*Fig. 7.10* Bird movements on the Severn estuary, in relation to the Severnside airport proposal

from their Flat Holm and Steep Holm roosting sites in the morning and return in the evening. Other groups of gulls centre their feeding activities on the Newport refuse tips and probably roost further up the estuary.

Although the existing airport at Rhoose is situated on the coast west of Cardiff, it has no significant gull problem because its approaches do not cross any major gull flight lines. The situation would be different if the Severnside airport were built. Aircraft would take off and land along the line of the estuary to reduce the noise disturbance over built-up areas. Bird movements in the vicinity of Flat Holm and Steep Holm would be unlikely to cause problems because the birds at this point seem to move at altitudes below 2500 ft (760 m) whereas the aircraft would be flying between 5000 and 28 000 feet (1680–8540 m) (Fig. 7.10*a*). In the vicinity of the airport however, gulls and aircraft would be flying at much the same level and problems could certainly arise (Fig. 7.10*b*). At present the low level gull movements along this part of the coast are centred on the refuse tips at Newport. If, as proposed, the airport reclamation scheme includes two large reservoirs (Fig. 7.10*c*), the gulls would almost certainly adopt these as a night-time roosting site and move every morning and evening across the airport runways. A possible solution would be to discontinue rubbish tipping at Newport. Even then the birds might simply transfer their feeding activities to the Cardiff refuse tips and retain the reservoirs as a roosting site, a course of action which would still result in regular flights across the airport.

## The ecological commentary

In a development area, ecology can clearly make a direct contribution to planning in assessing possible impacts on conservation sites, fisheries and crops. Equally important is the demonstration that ineffective urban waste disposal is likely to lead to eutrophic reservoirs and gull problems at airports,

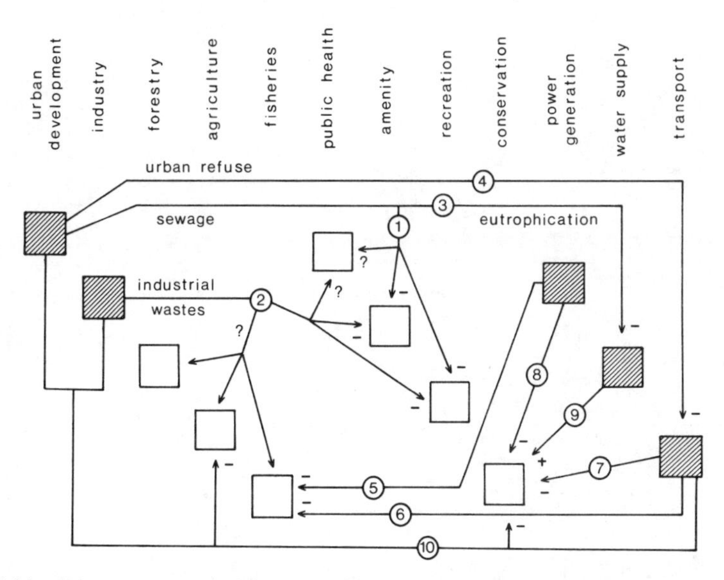


*Fig. 7.11* Diagram summarising the main issues included in an ecological commentary for the South Wales Coastal Development area. Shaded blocks represent the overall planning strategy: *1.* Effects of sewage discharges; *2.* Effects of industrial waste discharges; *3.* Eutrophication of reservoirs; *4.* Interference by gulls with airport operations; *5.* Effects of water extraction and tidal barrage on salmon fishery; *6–7.* Effect of Severnside airport on salmon fishery and wading-bird populations; *8.* Effect of tidal barrage on wading-bird populations; *9.* Creation of waterfowl habitats by reservoir construction; *10.* Encroachment on rural land; (−) = constraints; (+) = opportunities

two possibilities still largely ignored by engineers and planners. In the matter of assessing direct health hazards to man, the ecologist can only look at part of the picture and needs the collaboration of medical specialists.

For any region with a defined planning strategy, it is possible to summarise the likely conflicts of interest in the form shown on Fig. 7.11. The shaded blocks represent the planning strategy and the links show the relationships susceptible to some kind of ecological assessment. As will become apparent from the case histories which follow, the balance of the ecological contribution changes according to the type of area being considered.

# 8 A National Park

On a world-wide basis, most National Parks cater for some blend of wildlife conservation, landscape protection and public recreation. In Britain, however, greater emphasis is given to scenic amenity and recreation than to wildlife conservation. The British Parks are also atypical in that the land within them remains largely in private ownership.

The Brecon Beacons National Park in South Wales (Fig. 8.1) illustrates some of the problems of planning and management typical of Britain's Parks. Scenically the Park is dominated by the four plateaux of old red sandstone which form its principal mountain features. In addition it includes a wide diversity of wooded gorges, caves and waterfalls associated with the Carboniferous limestone and millstone grit areas along its southern margin. Apart from general sightseeing, the Park accommodates a variety of specialised recreational activities including angling, boating, caving and pony-trekking (Fig. 8.1). The main habitats of biological conservation interest are: the caves and woods on the carboniferous limestone, the large eutrophic lake at Llangorse, and a number of cliff sites supporting arctic-alpine plant communities near their southern limit in Britain.

Since its establishment in 1957, there has been a steady increase in the number of visitors using the Park. There is growing concern that this intense recreational pressure is detracting from scenic amenity, damaging biological conservation sites, and creating conflicts between different recreational groups. This pattern seems to be fairly typical of parks all over the world. In Britain, because the establishment of a park involves no significant change in land ownership, previously established land uses continue with little change. Inevitably some of these land uses are at variance with the ideals of the Park and need to be controlled to prevent them from becoming intrusive. Others present fewer problems and may indeed make a positive contribution to amenity and recreation. This need to find compromises between new and established land uses is the other major task in Park planning and will be examined first.

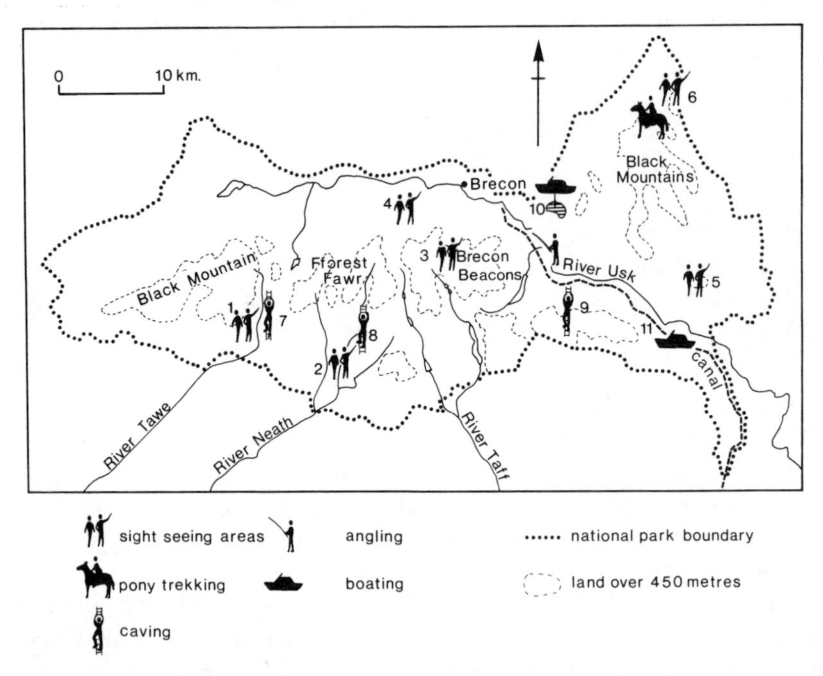


*Fig. 8.1* Map showing main scenic and recreational areas in the Brecon Beacons National Park

*1.* Dan yr Ogof show caves
*2.* Neath waterfalls
*3.* Brecon Beacons
*4.* Mountain Centre
*5.* Sugar Loaf
*6.* Hay Bluff
*7.* Upper Tawe Valley caves
*8.* Upper Neath Valley caves
*9.* Llangattock caves
*10.* Llangorse Lake
*11.* Brecon-Newport Canal

## Limestone quarrying

Although limestone quarrying is the only significant industrial activity in the Brecon Beacons National Park, it is potentially a very intrusive one. It is a source of dust and noise and can disfigure the landscape and damage conservation sites. These effects on biological conservation are of particular interest. Quarrying is associated with the carboniferous limestone outcrop which runs in a relatively narrow band across the southern part of the Park. At present there are six working quarries in this zone. Its biological interest is based partly on an extensive series of caves and partly on some scattered woodland remnants (Fig. 8.2).

### *Cave sites*

The biological interest of the caves centres on the unusual communities that have developed in these dark underground habitats which are devoid of

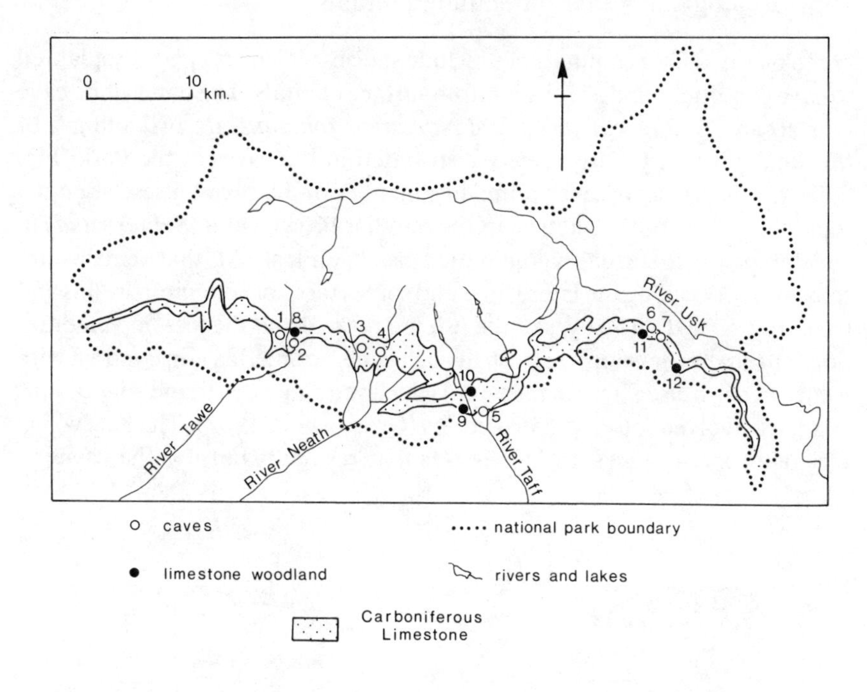


| *Caves* | *Asellus* | *Niphargus* | *Blanched trout* | *Horseshoe bats* | *Status* |
|---|---|---|---|---|---|
| *1.* Dan yr Ogof | × | × | | | part show cave |
| *2.* Ogof Ffynnon Ddu | × | × | × | | NNR |
| *3.* Little Neath River Cave | × | × | × | | |
| *4.* Porth yr Ogof | × | × | | | |
| *5.* Ogof y Ci | × | | × | | |
| *6.* Agen Allwedd | × | × | | × | NNR |
| *7.* Eglwys Faen | × | × | | × | NNR |

| *Woods* | *Ash dominant* | *Beech dominant* | *Endemic Sorbus spp.* | *Status* |
|---|---|---|---|---|
| *8.* Craig y Rhiwarth | × | | × | NTR |
| *9.* Penmoelallt | × | | × | FNR |
| *10.* Daren Fach | × | | × | NTR |
| *11.* Craig y Cilau | × | | × | NNR |
| *12.* Cwm Clydach | | × | | NNR |

*Fig. 8.2* Map and table of sites of biological conservation interest in the Carboniferous limestone zone.

(NNR = National Nature Reserve; FNR = Forest Nature Reserve; NTR = Naturalists' Trust Reserve) green plants. These communities include species which are specially adapted to cave conditions and are absent from surface habitats. The blind white cave crustaceans *Asellus cavaticus* and *Niphargus fontanus* are two animals of this kind which are fairly widely distributed in the caves of the Park (Fig. 8.3*e,f*). White trout are also found in a number of the caves. These are not a separate species but are a form of the familiar brown trout (*Salmo trutta*) in which most of the usual pigmentation has been lost. All that remains are some red spots along the lateral line and some traces of colour on the fins and back (Fig. 8.3*c,d*). The eyes of the fish are normal and the loss of pigmentation is probably developed afresh in each generation. This is in contrast with white cave fish elsewhere in the world which show more extreme adaptations and have evolved into separate species (Greenwood, 1967). The bats which are found in the caves are not permanent residents but use the caves as

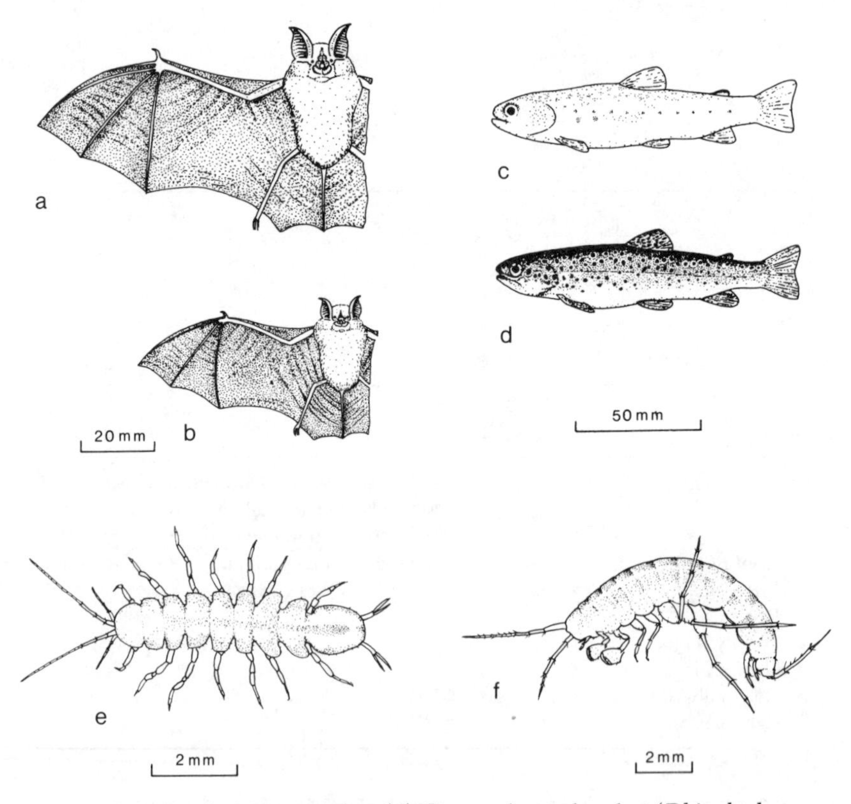


*Fig. 8.3* Some typical cave species: (*a*) Greater horseshoe bat (*Rhinolophus ferrumequinum*); (*b*) Lesser horseshoe bat (*Rhinolophus hipposideros*); (*c, d*) Blanched and normal brown trout (*Salmo trutta*); (*e*) *Asellus cavaticus*; (*f*) *Niphargus fontanus*

hibernation sites during the winter. The Agen Allwedd/Eglwys Faen series is a traditional overwintering site for horseshoe bats (Figs. 8.3*a*,*b*).

An unusual plant association found on the walls of some caves is the 'wall-fungus', which appears in the form of orange or white spangle-like colonies (Fig. 8.4). These colonies are made up of bacterial and fungal

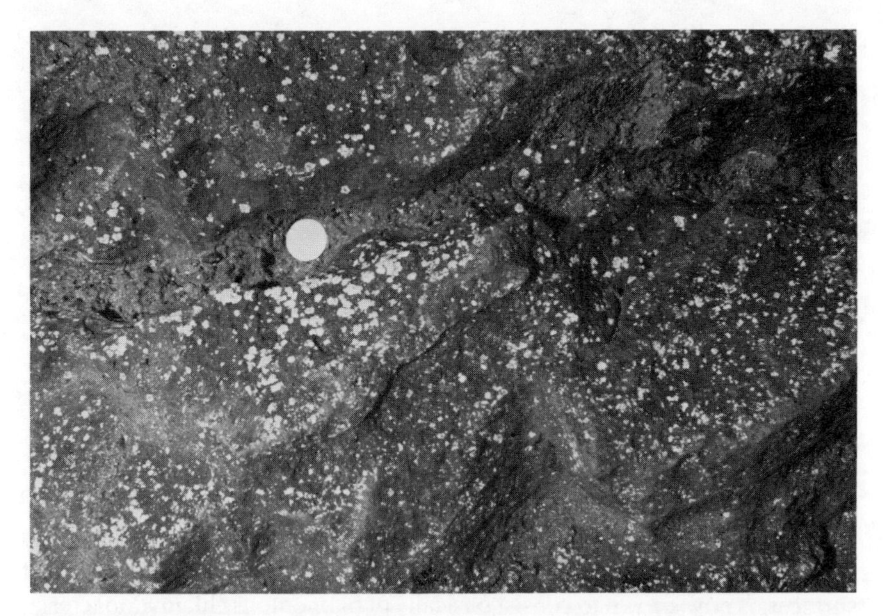

*Fig. 8.4* 'Wall fungus' in Ogof Ffynnon Ddu (scale shown by a $2\frac{1}{2}$p piece)

elements. Both the species involved, the bacterium *Streptomyces* sp. and the fungus *Fusarium* sp. are known from above-ground habitats, but in the combined form they are recorded only from caves. The relationship is almost certainly a response to the unusual nutritional circumstances in caves. The bacteria are able to use the organic matter entering the cave in percolation water and the fungus probably obtains its living from the tissues or waste products of its bacterial associate.

The absence of light from caves prevents the growth of green plants and, as a result, cave communities work differently from those at the surface (Williams, 1966). Cave food-chains are dependent on the limited amounts of organic material entering the cave from outside. This material comes in by two routes; some is contained in the water percolating through joints and cracks in the roof, the rest is carried in by streams. There is a widespread misconception that water entering caves in limestone is filtered clean by its passage through the strata. This is not the case, the fissures in limestone are quite wide enough to allow the passage of particulate material and this, together with dissolved organic matter, forms the primary nutrient supply in many cave pools (Fig. 8.5). Other pools usually at lower levels receive an input of organic debris as a result of flooding by the main streams in the cave. This debris, originating from the surface, includes such items as leaves, branches and even some algae and insect larvae.

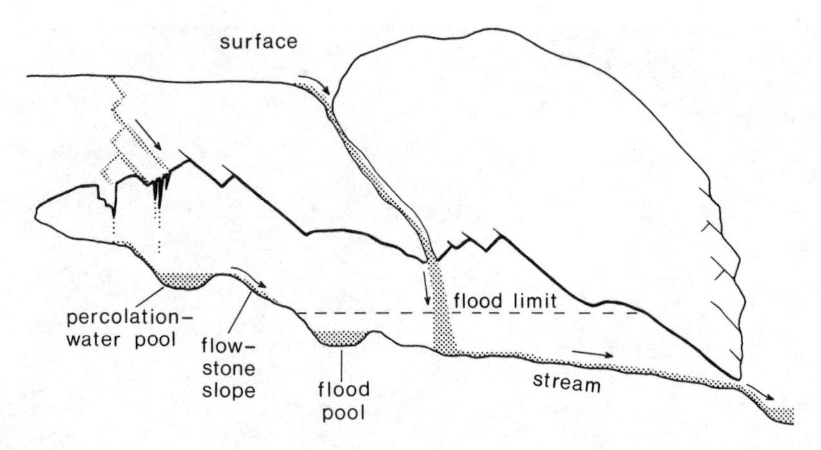


*Fig. 8.5* Aquatic habitats in caves

Bacteria are key organisms in the aquatic food chains of caves. The slimy coating, traditionally known as 'water fungus' which covers some wet surfaces, is now known to consist basically of filamentous chlamydobacteria. These bacteria use the organic matter brought in by floods and percolation water. The bacterial filaments in their turn trap particulate matter and it is this composite mass which provides some cave crustaceans with their main food supply. Measures devised for the protection of cave sites must have regard for these special characteristics of the whole community as well as the particular requirements of individual species.

The scientific interest of caves extends beyond their biological features in that they provide a commentary on the complex action of water on limestone strata. Some of their structural features such as stalactites, stalagmites, coloured flow-stones, and mud-flowers, also have a great aesthetic attraction, and parts of the Dan-yr-Ogof cave system have been developed as show caves for the general public. Caving as a sport is increasing in popularity in the Park and now involves both members of caving clubs and numerous groups from outdoor pursuits centres.

By destroying cave sites, limestone quarrying has the potential to interfere with all these interests. It might seem an easy matter to identify important cave sites and priority areas for quarrying and then arrange to keep the two interests separate. Even if this were administratively feasible there are a number of complications, many of them arising from the complexity of underground drainage systems. The cave systems in the Tawe Valley illustrate how underground water-courses bear little relationship to surface topography (Fig. 8.6). From surface features it is reasonable to assume that

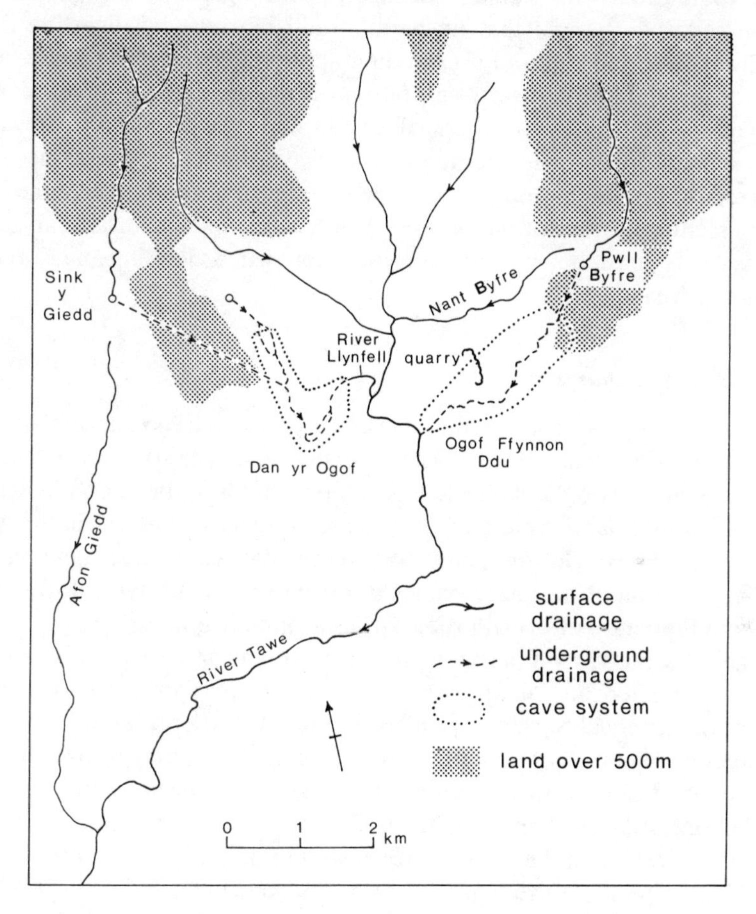


*Fig. 8.6* Map showing the relationship between surface and underground drainage in the Upper Tawe Valley

the water that sinks at Pwll Byfre re-emerges after a short underground passage to form the Nant Byfre. Similarly, the water that disappears at Sink y Giedd reappears as the Afon Giedd. Certainly these assumptions were made by the people who gave the sink points and the streams their names and have been perpetuated by the Ordnance Survey. In fact, when these underground water-courses are traced, using dyes, a very different picture emerges, the water which sinks at Pwll Byfre has no connection with the Nant Byfre. It passes through the Ogof Ffynnon Ddu cave system and discharges directly into the River Tawe some way downstream. Similarly the water from Sink y Giedd does not enter the Afon Giedd but flows in an easterly direction through Dan-yr-Ogof to reappear as the River Llynfell.

It would be easy for a quarry manager, operating in the vicinity of one of these sinks, to assume that no harm would be done by diverting these headwaters to some point lower in their apparent courses. In practice such an action would have disastrous consequences for the cave involved. In a similar way, oil spilt on a quarry floor can find its way into a cave water-course some miles from the quarry site. Cave drainage systems not only represent important supply routes for underground food chains, they are also essential to the formation and maintenance of structural features in caves. This must be a major consideration when devising measures to protect cave systems.

### *Limestone woodlands*

The woods on the Carboniferous limestone represent fragments of what was once a more extensive forest. There is good evidence to suggest that the natural climax woodland on the limestone would have been dominated by beech (*Fagus sylvatica*) in the east and ash (*Fraxinus excelsior*) in the west. (Fig. 8.2). This conclusion is based on present-day woodland fragments and the distribution of old place names referring to beech (Hyde, 1961). The western limitation of beech probably has a climatic basis.

The ash woods are of particular interest as many of them include unusual species of whitebeam (*Sorbus* spp). This genus of trees produces a very wide range of genetical variants, numbering about 100 species in the north temperate region. Some of them are extremely localised in distribution. *Sorbus leyana* for example occurs only on the limestone cliffs on either side of the Taff Valley at Penmoelallt and Daren Fach (Fig. 8.2, sites 9 and 10) and nowhere else in the world. *Sorbus minima* occurs only at Craig y Cilau (Fig. 8.2, site 11). Although botanists are somewhat divided about the conservation importance of these *Sorbus* species, some regard them merely as genetical curiosities, they are a feature of interest in these limestone woods and worthy of a measure of protection.

The overall planning problems generated by limestone quarrying are well illustrated by the area of the upper Tawe Valley shown in Figure 8.7. The quarry shown produces something like a third of a million tons of limestone a year for use as roadstone and flux for steel-making. As existing sectors of the quarry are exhausted, there is continual pressure to work adjacent areas of the limestone outcrop. However, the opening of new quarries on Craig-y-Rhiwarth raises objections on aesthetic grounds because this bluff is one of

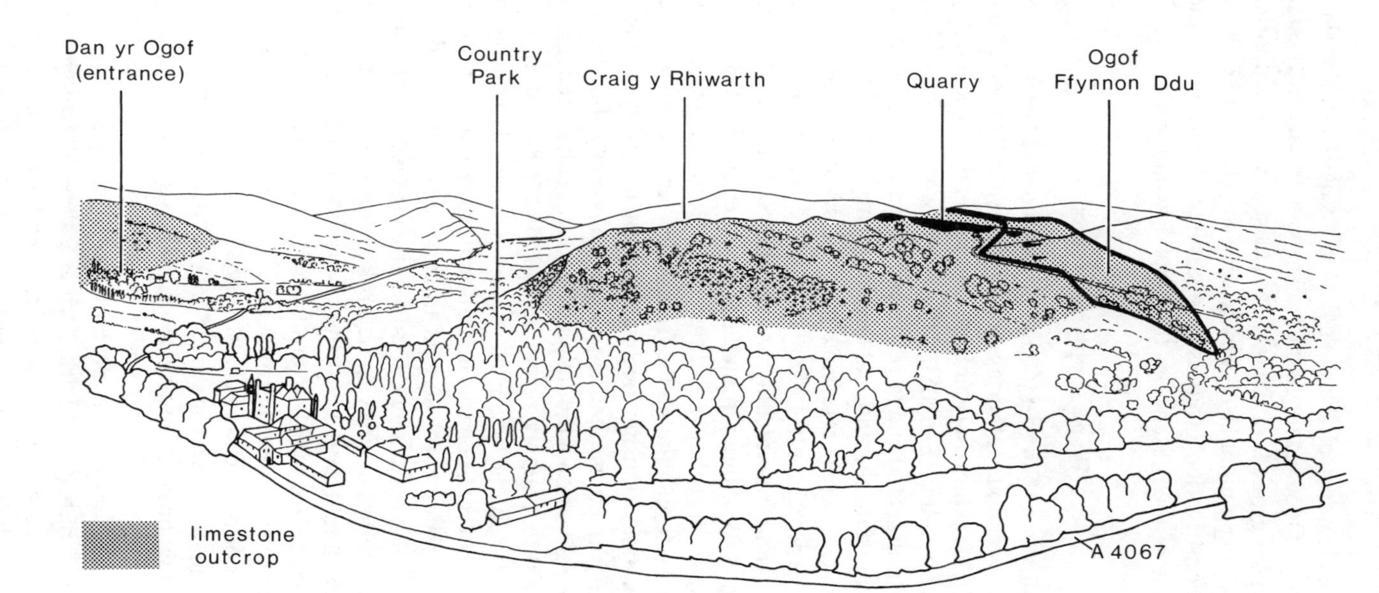


*Fig. 8.7* The Upper Tawe Valley

the major scenic features of the valley. It is visible from the show cave complex of Dan-yr-Ogof and forms the background to a recently established Country Park in the grounds of Craig-y-Nos Castle. It also supports a relict ash wood containing a number of *Sorbus* species. There are even stronger objections on conservation grounds to quarrying extensions in the other direction because of the presence of the Ogof-Ffynnon-Ddu cave system.

## Water supply reservoirs

The high rainfall, over 1524 mm (60 in.) per annum over a large part of the Park, makes it an obvious gathering-ground for water needed in industrial South Wales. Eight major reservoirs already exist in the Park and there have been proposals for others (Fig. 8.8). Although new reservoirs often involve the loss of agricultural land and terrestrial conservation sites, they can serve as a basis for important recreational developments.

Recreational activities on reservoirs have the advantage that they can be brought under stricter control than is possible on many natural lakes. Sailing, angling and bird-watching have been accommodated on the Park reservoirs (Fig. 8.8) without prejudice to water-supply operations and without creating conflicts between different recreational groups. The traditional policy of excluding the public from reservoir catchments because of the fear of contamination has become difficult to justify in the light of modern water-treatment methods and is gradually being abandoned here as elsewhere. One recreational limitation which must still be recognised is the unsuitability of direct-supply reservoirs for power-boat use, as spilt oil would seriously disrupt the biological processes essential to the operation of water-treatment filters. Even this limitation need not apply to 'regulating' reservoirs where water is transferred to extraction points and treatment works via a natural river channel. In these circumstances small amounts of spilled oil have the opportunity to be broken down and dispersed before reaching the treatment works. All the reservoirs in the Park function as direct-supply reservoirs and their unsuitability for power-boats and water skiing has concentrated these activities on Llangorse Lake where they have become a source of conflict and controversy.

Although the upland reservoirs in the Park serve as overwintering sites for wildfowl, their low productivity makes them less suitable for this purpose than the more productive reservoirs in the lowlands. The Cantref, Llwyn-onn and Pentwyn reservoirs support far fewer overwintering wildfowl than lowland reservoirs outside the Park, such as Eglwys Nunydd and Llandeg-fedd (Table 8.1). The apparent anomaly of the high counts at Talybont, a reservoir with an upland catchment, is explained by the fact that it serves as a day-time refuge for ducks from Llangorse Lake (Fig. 8.8). This lake is highly

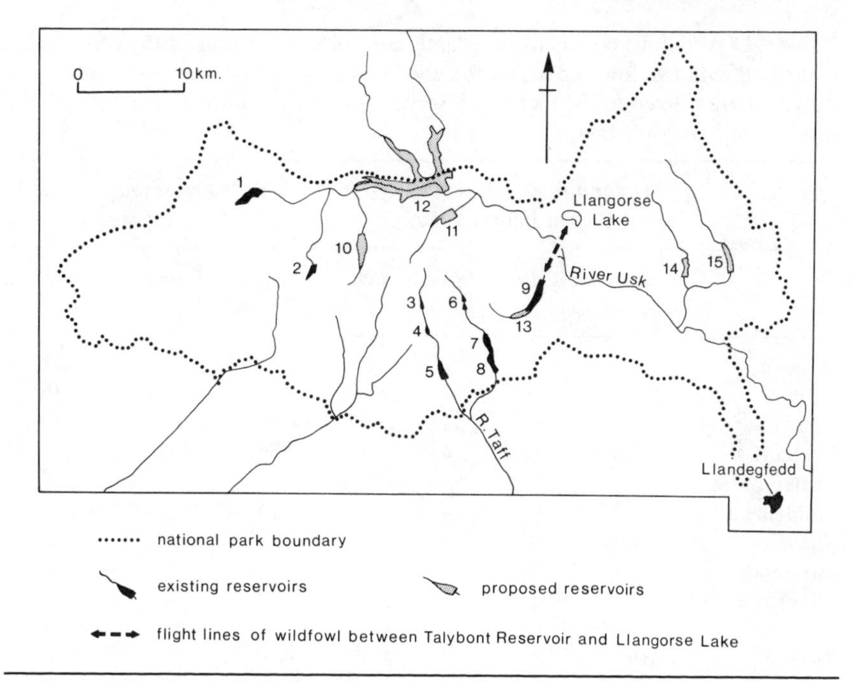


| | *Existing reservoirs* | | | *Proposed reservoirs* |
|---|---|---|---|---|
| | *Sailing* | *Angling* | *Bird watching* | |
| *1.* Usk | | × | | *10.* Senni |
| *2.* Cray | | × | | *11.* Tarell |
| *3.* Beacons | × | × | × | *12.* Big Usk |
| *4.* Cantref | | × | × | *13.* Talybont extension |
| *5.* Llwyn-onn | | × | × | *14.* Grwyne Fechan |
| *6.* Neuadd | | × | | *15.* Grwyne Fawr |
| *7.* Pentwyn | | × | × | |
| *8.* Pontsticill | × | × | | |
| *9.* Talybont | | × | × | |

*Fig. 8.8* Map and table of existing and proposed reservoirs in the Brecon Beacons National Park

productive and provides a wide range of feeding sites for ducks. However, winter water-skiers are a frequent source of disturbance and the birds tend to rest at Talybont during the day and return to feed at Llangorse during the night. Apart from representing habitats in their own right, reservoirs can also serve this important function of providing refuges for birds disturbed from heavily-used natural lake sites.

*Table 8.1* Wildfowl counts from upland reservoirs in the Brecon Beacons Park compared with two lowland reservoirs outside the Park. The high counts at Talybont are related to the fact that it serves as a refuge for ducks from the productive Llangorse Lake.

| | *Upland reservoirs in the Brecon Beacons Park* | | | | *Lowland reservoirs* | | |
|---|---|---|---|---|---|---|---|
| | *Beacons* | *Cantref* | *Llwyn-onn* | *Pentwyn* | *Eglwys Nunydd* | *Llandeg-fedd* | *Talybont* |
| Wigeon | – | – | – | 3 | 5 | 380 | 41 |
| Teal | 3 | – | – | 28 | 39 | 112 | 36 |
| Mallard | 3 | 3 | 83 | 30 | 14 | 594 | 239 |
| Pochard | 13 | 3 | 4 | 20 | 256 | 196 | 209 |
| Tufted duck | – | 3 | 3 | – | 140 | 64 | 52 |
| Goldeneye | – | 1 | 4 | 2 | 13 | 2 | 7 |
| Shoveler | – | – | – | – | 16 | – | 3 |
| Goosander | – | – | 1 | 15 | – | 1 | 19 |
| Other species | – | – | – | – | 7 | 2 | – |
| Total | 19 | 10 | 95 | 98 | 490 | 1351 | 606 |

(Counts represent annual maxima for each species averaged over a series of winters).

## Forestry and farming

From the evidence of buried-tree remains and the analysis of pollen grains in peat deposits, we have good reasons for thinking that the natural vegetation of the Park was broad-leaved woodland below about 600 m, with heathland on the hill tops. The present landscape of high open moorlands and patchworks of woods and fields on the lower slopes reflects the substantial diminution of woodland brought about by timber extraction and agriculture. In this sense forestry and farming have served to create a landscape which is both scenically attractive and easily accessible to the visitor. There are, however, a number of exceptions to this harmonious relationship. Hill-farm improvements usually lead to increased enclosure and more restricted access. Upland afforestation typically involves the spread of coniferous plantations which many people consider to be scenically intrusive. Various approaches have been developed to meet these difficulties. In the Brecon Beacons Park, as in other Parks in Britain, financial incentives can be given to farmers who make special provision for visitor access, and informal agreements have been made between Park planning authorities and the Forestry Commission on tree-planting patterns.

There are also claims that biological conservation interests are threatened by upland farm and forestry development. A fairly clear-cut example arises from the practice of sheltering sheep in woodlands during the winter. This prevents natural regeneration and in the long run leads to a further reduction in the total extent of broad-leaved woodland.

It is frequently claimed that the new coniferous plantations support only very impoverished animal communities and so must be regarded as a serious threat to wildlife conservation. As has been discussed earlier (p. 6) this is something of a half-truth. Taking song-birds as an example, although the variety of species is likely to be less, a coniferous wood can support a song-bird population which is equal in total number to that in a broad-leaved wood at the same stage of maturity. Moreover, an upland coniferous plantation is likely to support ten times as many birds as the moorland it replaced.

Most of the critics of coniferous woods as biological habitats are also vociferous in their condemnation of these plantations on aesthetic grounds and undoubtedly a large element of subjectivity has entered into the judgment of conservation implications. As will be seen, such failures to distinguish between aesthetic or amenity considerations and biological conservation issues are a constant source of confusion and difficulty in National Park planning.

## The effects of recreation

There is always a risk in a National Park that recreational developments reach a point where attractive and valuable features are damaged and where one recreational activity starts to interfere with another. Both these situations have some biological aspects.

### *Water-based recreation*

On a reservoir it can be a relatively simple matter to prevent conflicts between different recreational activities by some kind of zoning system. This is seen at the interconnecting Pentwyn and Pontsticill reservoirs where a potential conflict between naturalists and yachtsmen has been avoided by restricting winter-sailing to the lower basin, whilst retaining the upper basin as a wildfowl refuge. Zoning can be effective even with a much wider range of recreational activities, as is well illustrated at Llandegfedd Reservoir, just outside the Park (Fig. 8.9). Here the prohibition of winter-sailing prevents the disturbance of wildfowl. In summer the potential conflict between sailing and angling from the bank has been avoided by allocating areas for each activity. Definite areas have also been set aside for canoeing and diving. This

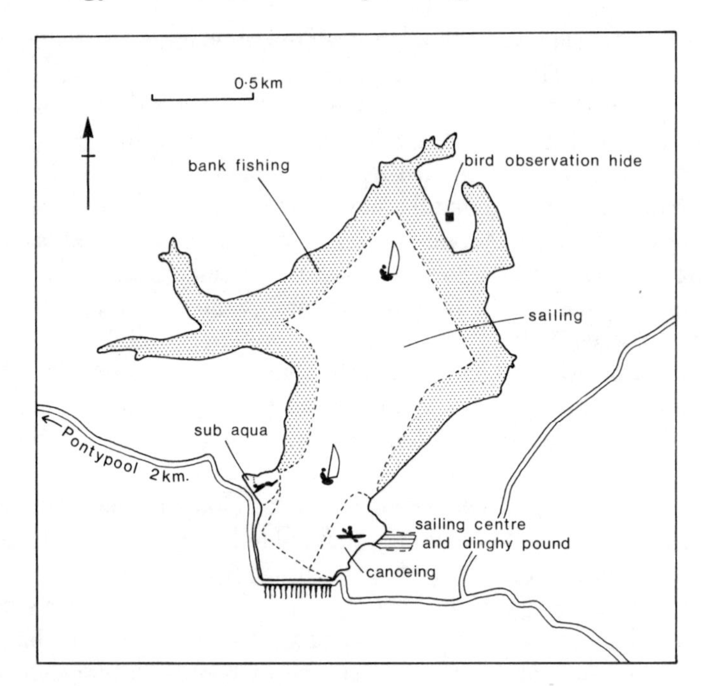


*Fig. 8.9* Zoning of recreational activities at Llandegfedd Reservoir

kind of centralised control is possible on reservoirs where there are no complications of ownership nor of established rights and traditions. It is more difficult to achieve on rivers and natural lakes and Llangorse Lake provides a particularly instructive example of the problems which can arise.

## *The recreational use of Llangorse Lake*

In summer Llangorse Lake supports a variety of recreational activities including water-skiing, canoeing, rowing, sailing and angling. Extensive camping and caravan sites have also been developed around the lake to accommodate holiday-makers (Fig. 8.10). The use of speed-boats either by themselves or towing water-skiers interferes with other activities on the lake. When speed-boats are operating, people in canoes, rowing-boats and sailing dinghies feel that they are denied the use of the centre of the lake because of the wash and the danger of collisions. Anglers similarly consider that the speed-boats prevent them from using long lines to 'troll' for pike.

Local naturalists, whose interest in the lake is based largely on its bird populations, are another group which has consistently regarded the speed-boats with disfavour. The standard work on the birds of the County (Ingram & Salmon, 1957) included a prediction that 'the growing commercialisation

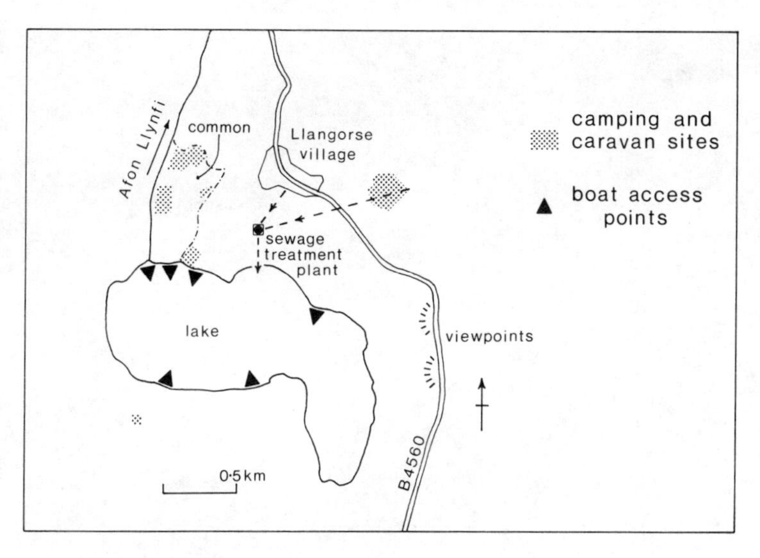


*Fig. 8.10* Llangorse Lake

of the lakeside, the advent of motor-boats and a sailing club, all tend to more and more uncontrolled disturbance which will soon reach a point at which the water-birdlife will be reduced to negligible proportions'. These writers talk of a reduction in the breeding numbers of great-crested grebes (*Podiceps cristatus*), sedge warblers (*Acrocephalus schoenobaenus*), and reed warblers (*A. scirpaceus*) at the lake by the 1950's and attribute it to recreational developments. More recently the appearance of algal blooms on the lake and the occasional massive mortality of fish (Fig. 8.11) have produced further objections to speed-boats. For example in July 1969 the Newsletter of the local Naturalists' Trust expressed the view that 'exhaust fumes, petrol and oil from motor-boats . . . are poisoning the lake fauna and destroying the balance of micro-organisms', adding further that 'the wash from speed-boats breaks up the reeds and submerged vegetation causing an accumulation of organic matter. This in turn causes a rapid build-up of certain bacteria which use up all the available oxygen resulting, on occasion, in the death of larger animals such as fish'. It was in this climate of opinion that by-laws were proposed in 1971 which would have amounted to a ban on speed-boats and water-skiing. These were, however, rejected at a public enquiry.

In spite of the widespread conviction that speed-boats are responsible for the unwelcome biological changes at the lake, at least some of these changes can be explained more plausibly in other ways. Llangorse is a shallow, well-mixed lake which suffers little deoxygenation in summer; this is not therefore a likely cause of fish mortality. However, the algae causing the

*Fig. 8.11* Fish mortality at Llangorse Lake; large numbers of roach (*Rutilus rutilus*) died in the summers of 1968 and 1970. (Photograph – Roy Adams)

blooms, *Aphanizomenon*, *Anacystis* (*Microcystis*) and *Anabaena* are all species which are known to be capable of killing fish by the release of toxic breakdown products. On other lakes and reservoirs these species are associated with nutrient enrichment and this is a likely explanation at Llangorse, where a sewage plant discharging into the lake has become progressively overloaded. The plant was probably adequate to serve the 360 people living in Llangorse village but in recent years it has received an additional input from up to 600 visitors using local camp sites. (Fig. 8.10).

As for adverse effects of speed-boats on birds, there is no firm evidence that these have actually occurred. In recent years, ten breeding pairs of great-crested grebes have commonly been recorded at the lake, together with an estimated 100 pairs of reed warblers and about half this number of sedge warblers. This suggests that there has been little overall change since the beginning of the century. The disturbance of ducks by water-skiers in winter is unlikely to have a permanent effect on overwintering populations because of the availability of Talybont Reservoir as a refuge. It is perhaps natural to assume that speed-boats which create so much disturbance in human terms must have equally drastic effects on plant and animal populations. However, this latter issue, like that of birds in coniferous plantations is not one which can be resolved on a subjective basis. A careful appraisal of the admittedly incomplete evidence strongly suggests that the biological changes at the lake are less likely to be checked by restrictive measures directed against speed-boats than by the construction of a new sewage plant with tertiary treatment facilities or the diversion of the sewage outfall into the outflow stream.

*Vegetation wear*

Like many areas under heavy recreational pressure, the most used sites in the Park show evidence of vegetation wear, as, for example, in the vicinity of the Mountain Centre, on the main approach path to the Brecon Beacons, and at vantage points such as Hay Bluff. The most widespread objections to vegetation wear are that it can lead to muddy footpaths and car parks and can produce areas of soil erosion which some people find unsightly. Some observers claim that the process is causing widespread damage to important biological systems. A typical example of this view is expressed by Ellis (1974) when describing vegetation wear on moorland near the Mountain Centre:

> 'the intense over-use by cars and visitors who have access to the open moorland here are seriously disturbing the ecological balance of the vegetation and small moorland wildlife, as well as initiating quite extensive soil erosion damage, all setting up a chain reaction which must lead to the destruction rather than the preservation of the local moorland ecosystem.'

Such statements greatly exaggerate the scale and significance of the changes. This is well illustrated by examining the process of vegetation wear on the paths approaching the summit of the Brecon Beacons. Away from the paths, bilberry (*Vaccinium myrtillus*), mat grass (*Nardus stricta*), and purple moor grass (*Molinia caerulea*) are the dominant species. Once a path has been started, these are replaced by more resistant plants such as sheep's fescue (*Festuca ovina*) and bent grass (*Agrostis tenuis*). With further pressure even these species disappear and the soil is exposed. The soil cover may then break up and, according to the local drainage pattern, be washed downhill or transformed into an area of mud. Many paths show the various stages of this sequence with the most extreme changes being apparent in the most used sections of the path (Fig. 8.12). A simple method of gauging the trampling pressure is to note the disturbance caused to lines of pliable wires inserted at intervals across the path (Bayfield, 1971).

The bilbury/mat grass/moor grass community, affected by this process is not the natural climax vegetation for the hills, it is an impoverished remnant of the species-rich heaths and woodlands which existed before the advent of sheep grazing. It is moreover a community extending over many thousands

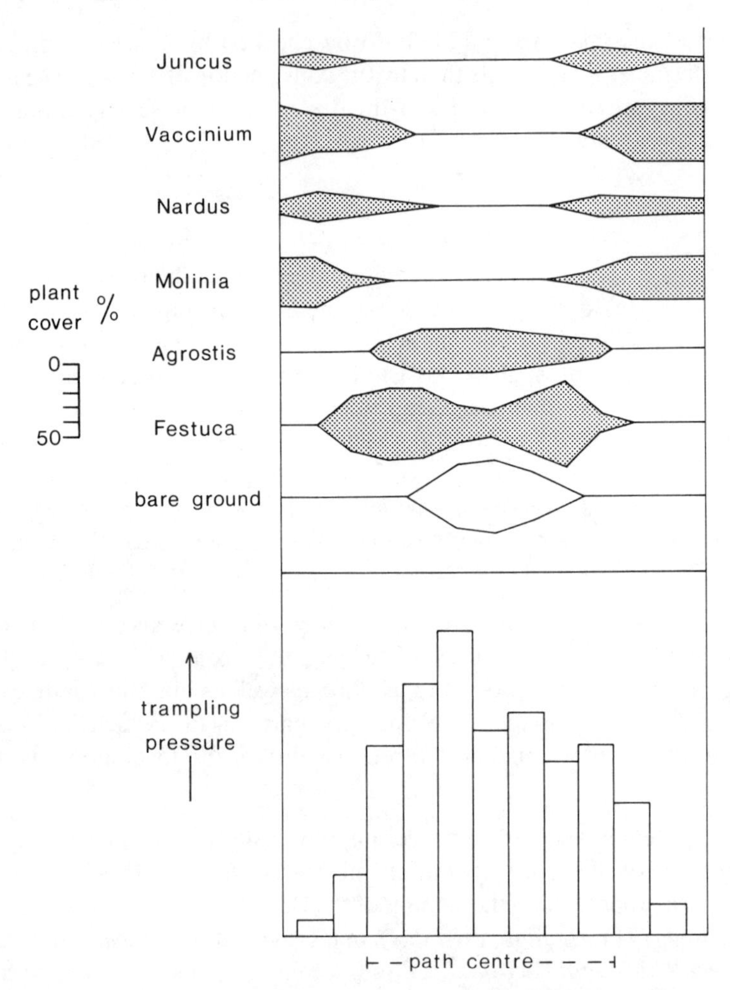


*Fig. 8.12* Vegetational changes caused by wear on the approach path to Pen y Fan in the Brecon Beacons (data collected by C. H. L. Green, University College, Cardiff)

of hectares in the Park and elsewhere in Britain. The disappearance of a minute fraction of this assemblage as a result of recreation is therefore of negligible conservation significance. The same is true of the other common moorland communities. Where no biological conservation issue is involved, it is appropriate to combat vegetation wear by whatever means are both practical and aesthetically acceptable. In the Park this has included the surfacing and rerouting of paths, the incorporation of nylon nets into car-parking surfaces, and the erection of bollards to prevent cars being driven onto open moorland.

Where the vegetation at risk is of conservation significance as it is for example on the approaches to Snowdon (p. 25), the options become more restricted and the strategy of diverting the public is usually the only one available. Arctic-alpine plant communities cannot be made hard wearing.

*Caving and conservation*

In caves, the conservation issues are more clear-cut and many of the important species in this habitat are liable to disturbance by cavers. The aquatic crustaceans are particularly vulnerable because many of the shallow-water sites they occupy are situated in passages which cavers use regularly. *Niphargus fontanus* usually occurs in pools fed by percolation water or occasional floods, whilst *Asellus cavaticus* is typical of water films, often those formed by overflow water from pools (Fig. 8.13). In these

*Fig. 8.13* A typical habitat for cave crustaceans, *Niphargus* occurs in the pools and *Asellus* on the flow stone slopes

situations *Asellus* is liable to be crushed by cavers' boots, *Niphargus* is affected when constant wear on the calcite rim of a pool causes the rim to fracture, releasing the water and making the habitat unsuitable. In a number of caves, pools have been sterilised unnecessarily by the dumping of rejected torch batteries and spent carbide from acetylene lamps.

In Porth yr Ogof, a cave used increasingly in recent years by outdoor pursuits groups, these processes have virtually eliminated *Asellus* and

*Niphargus* from any accessible parts of the cave. The two species have also declined over the past 20 years in the heavily used passages of the entrance series of Ogof Ffynnon Ddu, a part of the cave which receives about 3500 visitors a year. Fortunately, in most cave systems there are wet passages not used by cavers because they are too small to be negotiated or appear to lead nowhere. These places provide refuges for the animals (Fig. 8.14). However, the respite is not necessarily a permanent one because new routes are continually being developed by cave-explorers.

In addition to *Asellus* and *Niphargus* two other aquatic crustaceans occur in Ogof Ffynnon Ddu; one is the ostracod *Cypridopsis subterranea* which is known in Britain only from this cave, the other is the amphipod *Crangonyx subterraneus* which elsewhere is recorded only from some wells in Hampshire and Wiltshire. Very little information is available on their detailed distribution in the cave and some survey work is required to discover whether they too have natural refuges which protect them from cavers or whether special measures are necessary to ensure their survival.

Overwintering bats are the other vulnerable animals associated with caves. The Agen Allwedd and Eglwys Faen cave complex has been a traditional winter-roost for greater and lesser horseshoe bats (*Rhinolophus ferrumequinum* and *R. hipposideros*) but the numbers have declined as the

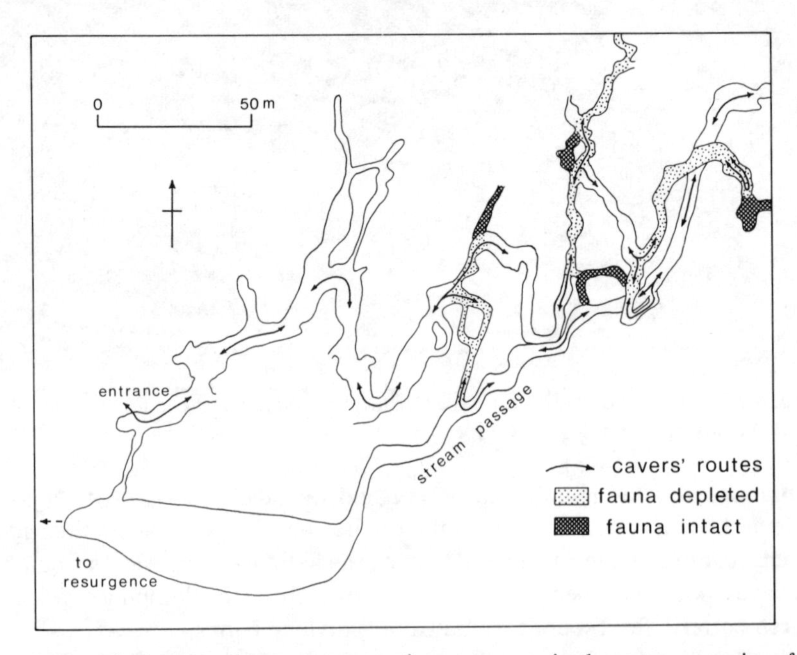


*Fig. 8.14* The impact of cavers on aquatic crustaceans in the entrance series of Ogof Ffynnon Ddu

cave has become more used. The greater horseshoe bat is a species which is reported to be declining generally in Britain and its numbers were estimated in 1971 to be as low as 500 individuals (Racey & Stebbins, 1972). The protection of this species in its cave roosts is thus a matter of some significance. Overwintering bats live on fat stores accumulated in the body from the previous summer. If a bat is disturbed and is active for a short period it uses up food stores which cannot be replaced. Repeated disturbances are likely to prevent it from surviving until the following spring. Intentional handling of bats by over-enthusiastic observers and ringers is one possible source of disturbance (Hooper, 1964; Stebbins, 1966) although not one which is important in the Park. Of more general significance is the fact that bats can be seriously disturbed, quite inadvertently, by the normal activities of cavers. Noise, fumes from carbide lamps, the use of small cooking stoves, and even the local increase in warmth caused by the presence of human bodies, can all have harmful effects on roosting bats.

Compared with other recreation/conservation conflicts, those in caves should be relatively easy to solve. Most cavers belong to caving clubs, and 'caving codes' which are at present principally concerned with safety matters, could easily be extended to cater for conservation. Moreover, where cave systems are designated as National Nature Reserves there is the opportunity through Nature Conservancy Council management committees to define and enforce conservation areas where these appear to be necessary.

**The ecological commentary for a National Park**

There is some similarity in the commentaries appropriate to industrial development areas and National Parks. In both situations there is a need to predict the likely effects of industrial and urban growth and the expansion of services such as roads and reservoirs. The important difference is that in the Park considerations of amenity, recreation and conservation are generally given priority (as is suggested by the shaded blocks in Fig. 8.15).

Ecological issues are frequently involved when attempting to limit the damage caused by extractive industries. Limestone quarrying is particularly troublesome in relation to the protection of caves. It does however avoid some of the problems associated with other industrial operations because of the relatively small amount of spoil produced. The china-clay areas of the Dartmoor National Park and the slate quarries of Snowdonia present far greater problems of rehabilitation and revegetation. Old limestone quarries also have the advantage that they can be used for climbing if loose rock is first stripped from the face.

Apart from the inundation of agricultural land and the occasional loss of conservation sites, reservoirs can represent an asset in a National Park

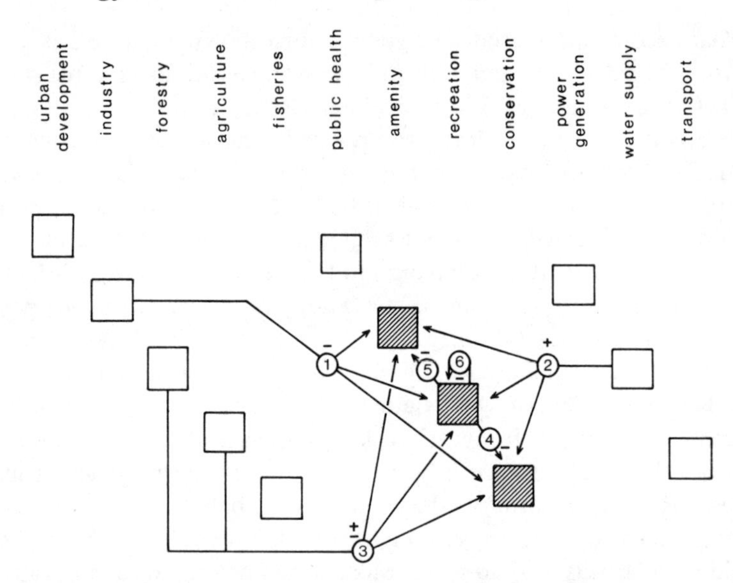


*Fig. 8.15* The main ecological issues involved in the development and management of the Brecon Beacons National Park: *1.* Effects of limestone quarrying; *2.* Effects of reservoir development; *3.* Effects of forestry and farming; *4.* Damage to conservation sites caused by recreational activities, e.g. caving; *5.* Detrimental effects of recreation on landscape quality e.g. vegetation wear and erosion; *6.* Interference between different recreational activities e.g. boating and angling

although sometimes the additional angling opportunities provided in the reservoir are offset by interference caused to migratory fish.

The relationship between forestry and farming activities on the one side and recreation, amenity and conservation on the other, involves a complex mixture of conflicts and opportunities. In this field and on the question of recreational impacts, the most useful contribution an ecologist can make is to bring some objectivity to the analysis of supposed biological changes. It is frequently assumed that amenity and conservation interests march together, that if coniferous plantations, speed-boats and worn footpaths are aesthetically objectionable, then they are also likely to have damaging effects on valued biological communities. This is not necessarily the case. Groups committed to landscape preservation have something to gain by blurring the distinction between amenity and biological conservation issues; it often allows amenity arguments to be apparently reinforced by scientific evidence. However, if in the long-run the planner is to make the right decisions, he needs to distinguish clearly between those situations which are concerned principally with amenity, those concerned principally with biological conservation and those which are genuinely a combination of both.

# 9 A derelict industrial area

Regions of extensive industrial dereliction are common in those developed countries where rapid and uncontrolled growth took place in the 19th century and where industrial activities were subsequently brought to a halt by depletion of raw materials or by market changes. It is now widely recognised that ecological information is of considerable value in the rehabilitation and redevelopment of such areas.

The coalfield valleys of South Wales (Fig. 7.1, p. 146) provide numerous examples of industrial derelication. Large areas of the lower Neath and Swansea Valleys are covered by furnace slags associated with non-ferrous metal smelting and slag tips from the old iron foundries remain a feature of towns at the north-eastern edge of the coalfield. The impact of the coal industry has been felt on a wider scale and in the valleys where coal mining has now ceased, there are major problems of rehabilitation and redevelopment. A typical area of this kind, which has been the subject of a detailed study, is the Afan Valley (Figs. 9.1, 9.2). Because of the steep valley sides, urban growth during the industrial revolution was concentrated on the lower slopes and valley floor, so too was the development of the roads, railways and trunk sewer system.

The early coal mines took the form of 'drifts' or 'levels' driven horizontally or at a shallow angle into the valley sides to exploit the upper coal seams. Later, shaft mines were sunk into the valley floor to work the deeper seams of steam coal and anthracite (Fig. 9.3). The coal industry dominated the valley from about 1800 to 1920, but since then it has declined steadily. The last major mine closed in 1970 and by 1973 only a single small drift mine remained. Dereliction from the coal industry is still very much in evidence, both in the form of spoil tips and in polluted discharges from flooded mines.

The other striking change which has taken place in the valley in the last 30 years has been the progressive replacement of hill farmland by forestry plantations. A number of the farm holdings have disappeared and others have been greatly reduced in size.. The reduction of farmland started when land on the valley floor was utilised for industrial and urban expansion. It

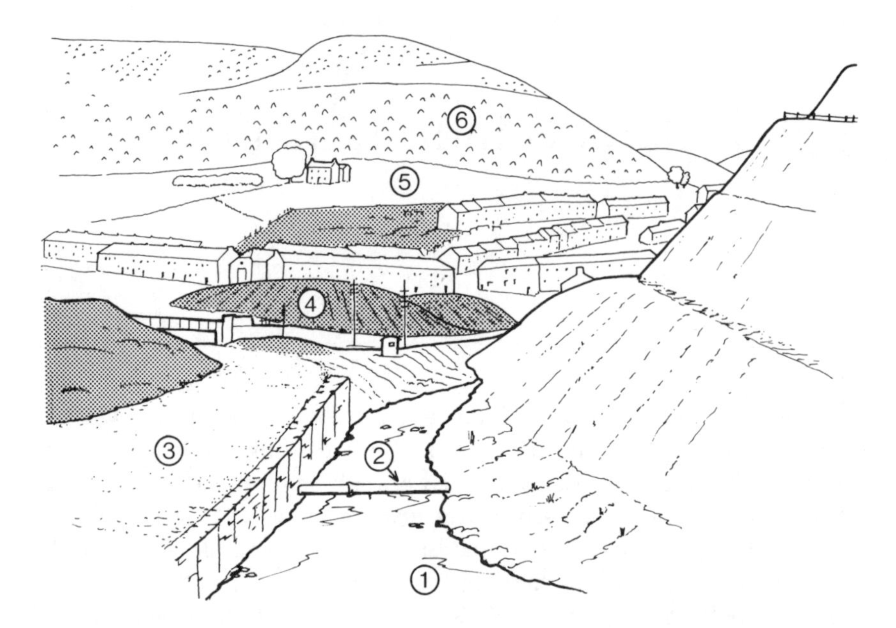


*Fig. 9.1* A typical coalfield valley town (Abergwynfi): *1.* River Afan; *2.* Trunk sewer; *3.* Abandoned railway; *4.* Coal and quarry spoil (shaded); *5.* Hill farm; *6.* Forest plantation

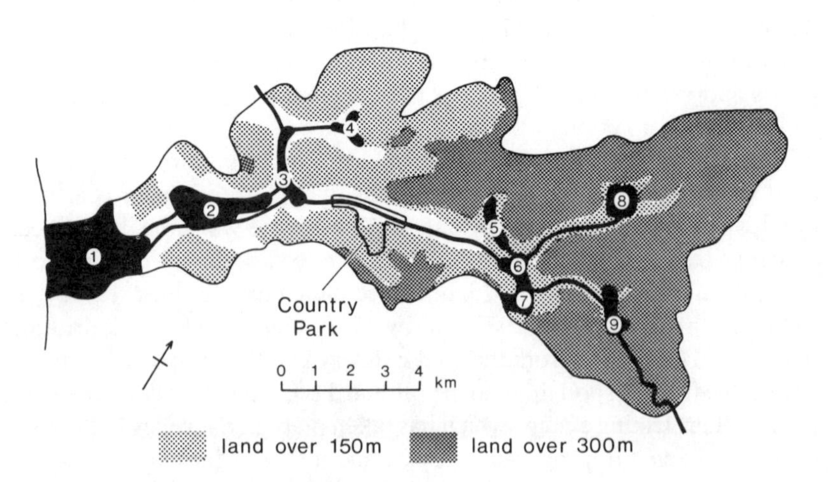


*Fig. 9.2* The Afan Valley showing relief, towns, roads and Country Park: 1. Port Talbot; *2.* Cwmavon; *3.* Pontrhydyfen; *4.* Tonmawr; *5.* Abercregan; *6.* Cymmer; *7.* Croeserw; *8.* Glyncorrwg; *9.* Abergwynfi

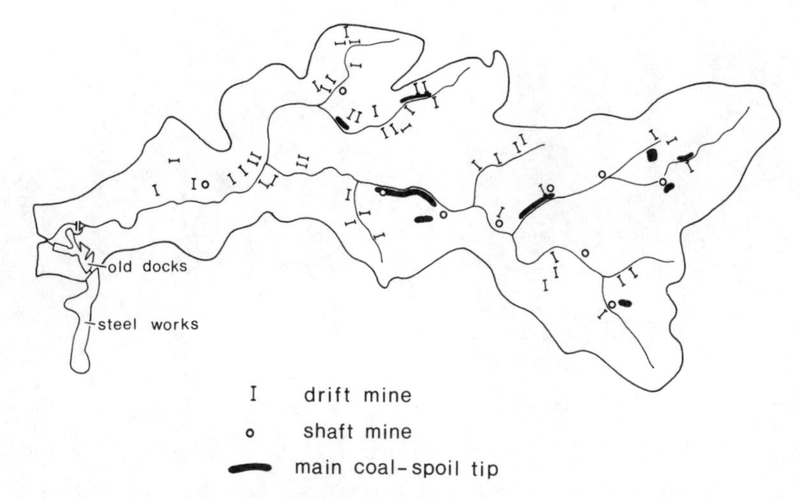


*Fig. 9.3* The maximum extent of the coal industry in the Afan Valley

continued when areas on the valley sides and plateau top were relinquished for afforestation. A farm which is typical of the area nowadays, bounded by the town below and a forestry plantation above, is shown in Fig. 9.1.

In these coalfield valleys, poor communications and a shortage of flat land for factory sites have greatly reduced the prospect of new industries being developed *in situ* to replace the coal industry and a large proportion of the working population now commutes to the coastal industrial area. This has led to planning strategies which lay special emphasis on the clearance of dereliction, the improvement of amenity, and in places where extensive rural areas have survived, the development of recreational facilities. The Afan Valley has the added interest of including a specific recreational development in the form of a Country Park.

## Amenity improvements

There has been a general appreciation of the need for amenity improvements in the towns of the coalfield valleys (Civic Trust, 1965; Welsh Office, 1967; Building Design Partnership, 1970). However, many of the efforts in this direction, either by the Local Authority or by individual householders, encounter interference from an unexpected quarter.

Every town in the Afan Valley (with the exception of Port Talbot) has its complement of straying sheep. Although some of the problems created by these animals are obvious, others turn out to be more subtle. Sheep cause widespread damage to gardens and many householders have supplementary barricades to their original fences in an effort to keep the animals out. (Fig. 9.4*a*). Landscaped areas, parks and even cemeteries need to be similarly

a

b

*Fig. 9.4* (*a*) A garden with special protection against sheep; (*b*) An unprotected garden

protected if they are to escape damage. For the householder an alternative strategy is to capitulate and accept that sheep will roam freely through the garden (Fig. 9.4*b*). Other valleys have their own variants of the gardening/animal straying problem. Some new housing estates in the eastern valleys of the coalfield have been designed to exclude straying ponies by means of a combination of walls, fences, and animal grids. Pedestrians can enter these estates only through special pony-proof swing gates. Unfortunately some ponies have now learnt to negotiate the grids by rolling over them.

In the Afan Valley refuse collection days have a special significance for straying sheep. On these days they move through the streets scavenging in dustbins and other rubbish containers, knocking off insecure lids, and where possible, overturning them (Fig. 9.5). In some towns as many as a 100 stray sheep live for long periods within the town boundary. Their regular feeding grounds include gardens and landscaped areas, areas of rough grass including derelict coal-spoil tips and tips of domestic rubbish, whilst on dustbin days a proportion of the animals turn their attention to dustbin scraps. Although to the outsider the straying sheep are a source of interest and amusement, to the local resident they represent damaged gardens and parks, scattered domestic refuse and soiled pavements. In these rather obvious ways they hamper efforts at environmental improvement. Additionally they

*Fig. 9.5* Sheep foraging amongst domestic rubbish in Abergwynfi

present a hazard to motor traffic. For example, in another coalfield area, the Borough of Merthyr Tydfil for which statistics are available, there were 275 reported accidents involving animals during the period 1969–1971, and farmers claimed that 1258 animals had been killed.

*Sheep-straying in relation to public health*

Some of the effects of sheep-straying are less obvious yet one at least warrants serious attention on public health grounds. Sheep in the South Wales valley towns are implicated in the transmission cycle of a tapeworm *Echinococcus granulosus* which can cause serious hydatid cyst infections in man (Echinococcosis). Figure 9.6 shows the life-cycle of this parasite. The adult tapeworm lives in dogs and the eggs of the parasite are passed out in dogs' faeces. The next stage of the cycle is completed when sheep pick up the eggs whilst grazing on contaminated herbage. The parasite develops in the sheep to form a large bladder-like structure, or hydatid cyst, in the liver or the lung. The life-cycle is completed where dogs have the opportunity to feed on infected sheep carcasses. Unfortunately, the life-cycle can be diverted so that the cyst develops in man instead of the sheep. This usually comes about when humans, particularly children, transfer eggs of the parasite from the fur of infected dogs to their hands, and eventually come to

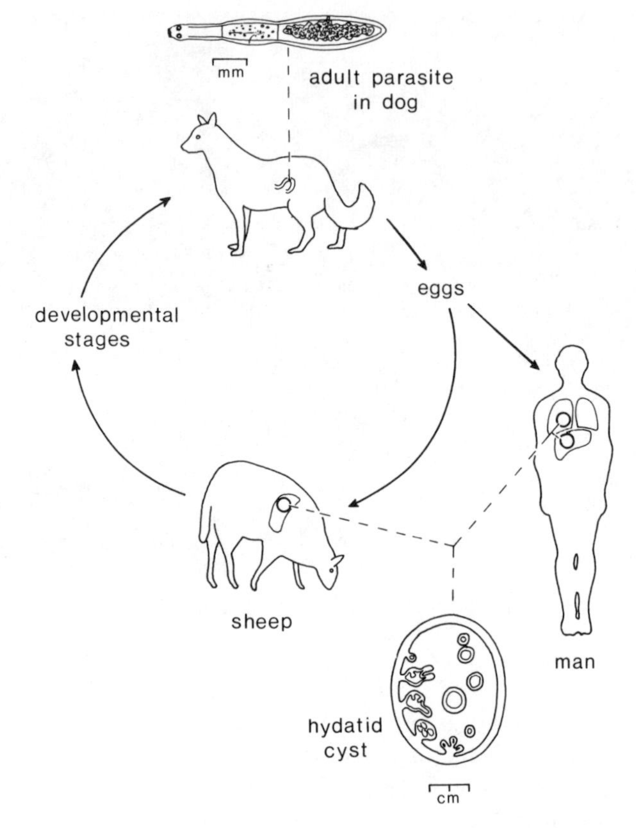


*Fig. 9.6* The life cycle of the tapeworm *Echinococcus granulosus*, which can cause hydatid cyst infections in man

swallow these eggs. The cyst may grow for between five and twenty years before it is diagnosed.

The incidence of the parasite is generally low in Britain compared with some other parts of the world. Nevertheless, the industrial valleys of South Wales, and, to a lesser extent, the North East Wales industrial zone, are recognised as important foci. The maps (Fig. 9.7) show the incidence of hydatid cases for the period 1947–1958 and illustrate the association of the infection with urban centres. No comprehensive figures are available for recent years but the incidence of hydatid cysts in sheep carcasses at slaughter houses in South Wales suggests that the parasite is still widespread. The close contacts between straying sheep, domestic dogs, and people in the valley towns of South Wales appear to create particularly favourable conditions for the transmission of the parasite. Cook & Crewe (1963) have pointed out how dogs in these areas have numerous opportunities to feed on

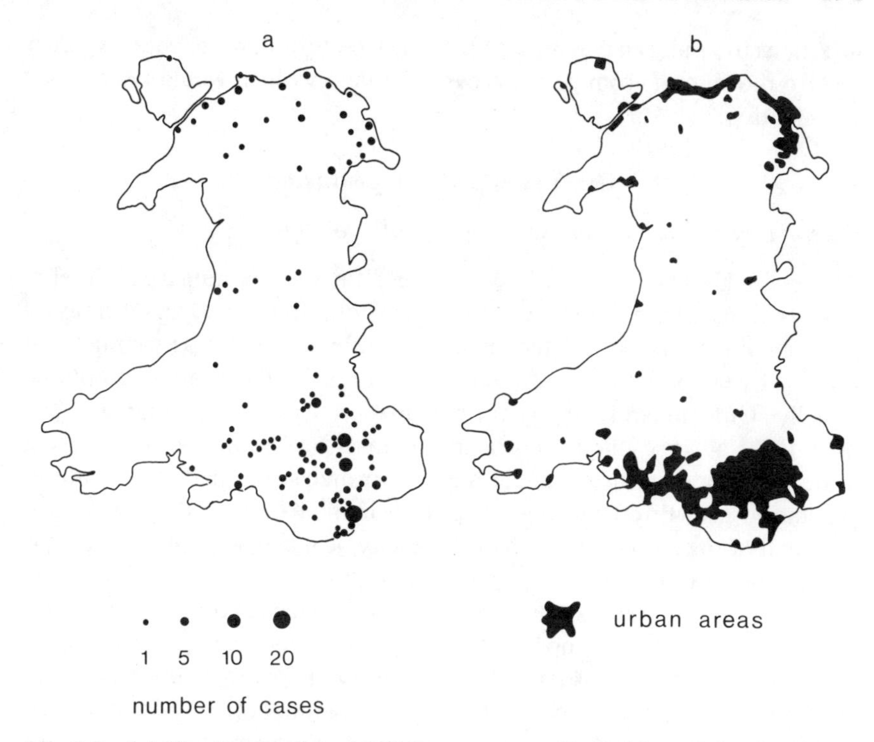


*Fig. 9.7* (*a*) The distribution of 193 human cases of hydatid cyst infections in Wales (1947–1958) (North Wales records from Jonathan (1960); South Wales records from Sully Hospital); (*b*) Urban areas in Wales (regions where the population density exceeded 1 person/acre in the 1951 census)

the carcasses of sheep, especially those knocked down on the roads or killed as a result of worrying by the dogs themselves. This ready access to sheep carcasses favours infection of dogs by the parasite.

The ultimate answer to this problem and to other difficulties associated with sheep is to keep the animals out of the towns and off major roads. The prospects of achieving this are discussed later. An immediate practical solution is for dog owners to try to ensure that their pets are kept free of the adult tapeworms. A number of suitable preparations are available from veterinary surgeons for this purpose. Those containing arecoline hydrobromide, bunamidine hydrochloride or Anthelin are commonly used against *Echinococcus* (Blood *et al.*, 1968). The need for children to wash their hands in between handling their pets and eating food is particularly important in situations where the pet could be harbouring the parasite.

The contamination of paved areas with sheep droppings is another aspect of urban sheep straying which has possible health implications, albeit of a minor nature. These droppings are a possible source of mild enteric infections for children playing either in the streets or in water courses which receive the run-off from paved areas, a situation which would bear closer examination.

## Other effects of sheep-straying

### *Sheep-straying and the landscaping of coal-spoil tips*

The coalfield valleys in South Wales are disfigured by numerous derelict coal-spoil tips, some of them over 100 years old. In recent years increasing attention has been given to techniques for reshaping and landscaping these tips, partly to render them safe and partly to make them more acceptable visually. Unfortunately, many tip reclamation schemes have ignored the potentially adverse effect of straying animals on the operation. Leaving aside for a moment artificial reclamation projects, the pattern of natural vegetation growth on the tips gives a clear indication of the destructive potential of the strays. In South Wales the typical course of plant colonisation on a newly abandoned tip is for the gradual establishment of a patchy grass cover. This rarely extends over the entire surface of the tip because it is continually being broken up by erosion and by the movement of the surface layers of shale. It is unusual to find tips in South Wales where trees have established themselves. Moreover, when an examination is made of the few tree-covered tips which do exist, it becomes apparent that in each case barriers, such as the river or a railway line, excluded animals from the tip whilst the trees were becoming established. Where sheep have access to a tip, and this is the usual situation, they prevent tree development by biting off the growing points of the young trees. With many tips of course, the infertile and unstable nature of the tip material is also an impediment to tree growth. There was a lesson here in practical landscaping but unfortunately few people realised it.

Where the intention is simply to reshape a tip and cover it with grass, straying sheep present no great problems, indeed they may serve a useful purpose. The grass mixtures usually applied consist of good herbage species including rye grass (*Lolium perenne*) which are much more attractive to sheep than the bent and fescue grasses on adjacent rough grazings. The large numbers of sheep which migrate on to unfenced newly-grassed tips do little harm to the grass cover and can save the local authority the expense of mowing.

Difficulties arise when the landscaping scheme involves tree planting in addition to grassing on an unfenced tip. Any trees less than about one and a half metres tall when planted, are likely to suffer serious damage from animal browsing. The twigs and foliage are the main targets for attack. In one tip-rehabilitation scheme near Ebbw Vale, approximately 80 000 small coniferous and broad-leaved trees were planted in 1971 in an attempt to create a landscaped area combining large copses with grassland. These unfenced tree plots attracted the attention of animals which caused extensive damage. In some areas the main shoots of 90 per cent of the trees were removed by animals (Fig. 9.8*a*). In spite of these initial failures, directly attributable to animal damage, and in spite of the dissemination of information about them, unfenced tree plots were still being set up in South Wales in 1973 with the same predictable results.

a

b

*Fig. 9.8* (*a*) Animal damage to unfenced tree plots on a landscaped tip at Ebbw Vale; (*b*) Sheep browsing on young Norway spruce trees in the Afan Valley

The point was, however, well taken by Glamorgan County Council and in all their major reclamation projects small trees have been successfully protected individually or by a fence around a group of trees. Although the animal browsing effect was revealed as part of the background to the Afan investigation, the tip reclamation schemes in the valley have not so far involved major tree-planting projects. Here the existing Forestry Commission fences give some measure of protection and birch trees have successfully established themselves on some of the tips inside the fences.

*Effects on oak woodlands and forestry plantations*

The Afan Valley has no great variety of sites of biological conservation interest. Probably the most important ones are the few small fragments of sessile oak (*Quercus petraea*) woodland. These are modified remnants of a formerly extensive oak forest in the coalfield. It might seem a simple matter to retain the Afan woodlands for conservation and amenity purposes; nonetheless, it was found that the pervasive effects of straying sheep extended to these habitats too. In woods that are freely accessible to sheep, few of the developing tree seedlings, which appear mainly in clearings, survive the grazing activities of the animals. Thus no natural replacements will survive to ensure continuity of the woodland unless fences are erected to protect developing seedlings.

Straying animals also gain access to forestry plantations where they do considerable damage to young trees (Fig. 9.8*b*). In the Afan Valley it was found that the main areas of sheep damage all involved Norway spruce (*Picea abies*) and sitka spruce (*Picea sitchensis*) in plantations less than 15 years old. The two species of spruce appear to be preferred to pine of a similar age. The damage takes place in spite of the fact that all the plantation areas have been fenced. The animals gain entry through breaches in the fence. Some of these breaches occur as the result of normal deterioration, but purposeful human interference is not unknown.

Some of the impacts of straying sheep in the Afan Valley are summarised in Fig. 9.9. In this valley and elsewhere in the coalfield the animals interfere with a wide range of interests and represent a serious impediment to any planning strategy which emphasises amenity, recreation and conservation.

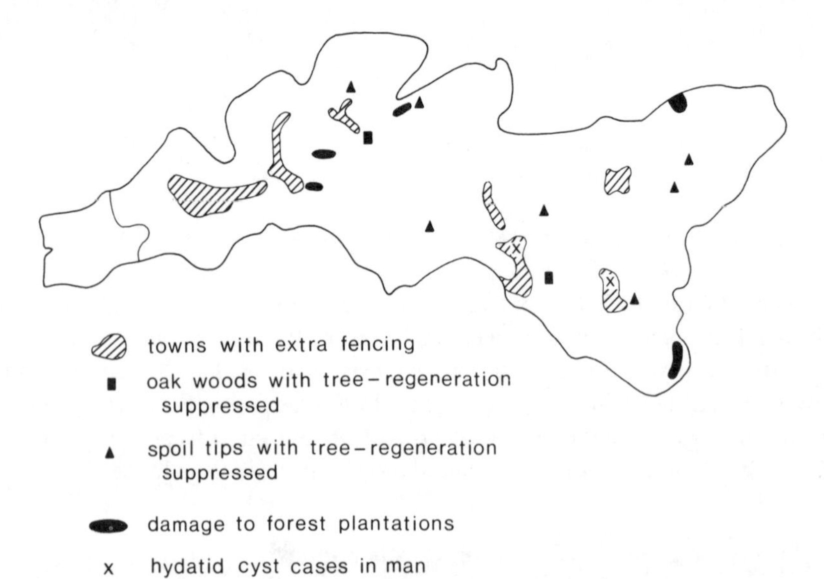


*Fig. 9.9* Map summarising the main consequences of sheep-straying in the Afan Valley

**Origins of the sheep-straying problem**

Sheep-straying has its origins in the industrial and urban development of the valley. During the period of industrial expansion large sections of the valley floor and lower slopes were used for housing, mine buildings, spoil heaps and road and railway development. This land thus became unavailable for farming and many farm units were so altered in structure as to become uneconomic. Further parcels of farm land were sold for afforestation (Fig. 9.10) and whilst this often provided the farmer with some much-needed

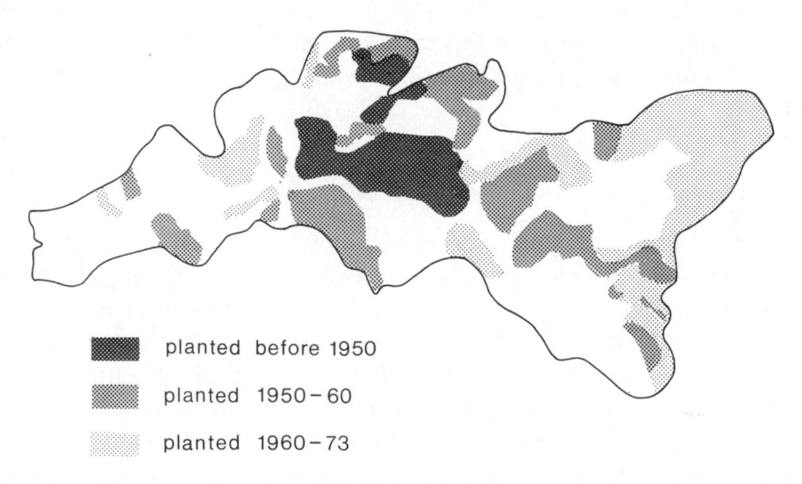


*Fig. 9.10* The expansion of Forestry Commission plantations in the Afan Valley

capital, it tended to weaken still further the working structure of the farm. Some farms have been reduced in this way to a few fields. Such units cannot in themselves support sufficiently large flocks to produce a significant financial return, unless the animals are able to find food in adjacent habitats such as forest plantations, towns and tips.

This would seem to be the basic cause of the animal-straying problem. The owner of a farm which has shrunk below a minimum size has a vested interest in the continuance of straying as a means of feeding his flock. Of course, not all farms in the valley are of this type. Some are viable units in themselves and for the owners of these farms the inconvenience of retrieving straying animals far outweighs the extra feeding areas accessible to them. This background needs to be borne in mind when seeking a solution.

Stock-proof fences are obviously one possibility. A fence can be used at a number of levels; on the small scale to protect a garden or a group of ornamental trees on a tip; on a large scale to protect a forest plantation or an entire housing estate. The weakness of large-scale fencing schemes is that they can be rendered ineffective by a single breach. It might seem more logical to fence the sheep into the farms rather than fence them out of everywhere else. However, the farmer who is short of land has no interest in restricting his sheep to his own land and denying them the advantages of straying. The farmer with a viable farm unit usually recognises the advantages of a ring fence around his farm but resists any suggestion that he should bear the cost of alleviating, by fencing, a situation which he considers is not of his own making. One solution which is being tried is for the local authority to impound straying sheep and return them to the farmer only on the payment of a fine. In the Afan Valley, this kind of scheme is currently meeting with some success in Glyncorrwg. Some combination of impounding and fencing would seem to offer the best prospect of alleviating the problem.

### *The balance between farming and forestry*

If, as seems to be the case, the sheep-straying problem has been aggravated by the transfer of farmland to forestry, it becomes appropriate to examine the ecological background to this change. To the traditional hill farmer in the rural parts of Wales a viable farm unit would consist of a combination of valley bottom fields, improvable slope land and rough hill-top grazings (Fig. 9.11*a*). In these terms, a coalfield valley farm which had lost some of its fields on the lower slopes to industrial or urban development would be regarded as a difficult proposition, principally because of the difficulties of improving the steeper slopes and the hill tops (Fig. 9.11*b*). A large proportion of the valley sides have slopes which exceed 13° and 20° which are the usually quoted working limits for ordinary and caterpillar tractors (Curtis *et al.*, 1965). The hill-top grazings in the coalfield are largely dominated by moor grass (*Molinia*) growing on waterlogged, peaty, gleyed soils, a situation which is also traditionally regarded as offering little scope for improvement. Such a poor prognosis for farming coupled with the claim that these difficult sites could be utilised successfully for forestry led to the progressive acquisition of land for afforestation. The land was transferred to forestry on a piecemeal basis, seemingly on the assumption that the change-over would eventually be complete.

As it transpired, a number of these assumptions were unsound. The steep slopes and hill tops which were regarded as so unpromising for farming have also proved to pose problems for the forester. On the plateaux tops the waterlogging of the soil near the surface prevents deep root development and as the trees grow taller a considerable number are blown over by the wind, a phenomenon known as 'windthrow' (Fig. 9.12). There have also been widespread failures in plantations on steep slopes in the upper parts of

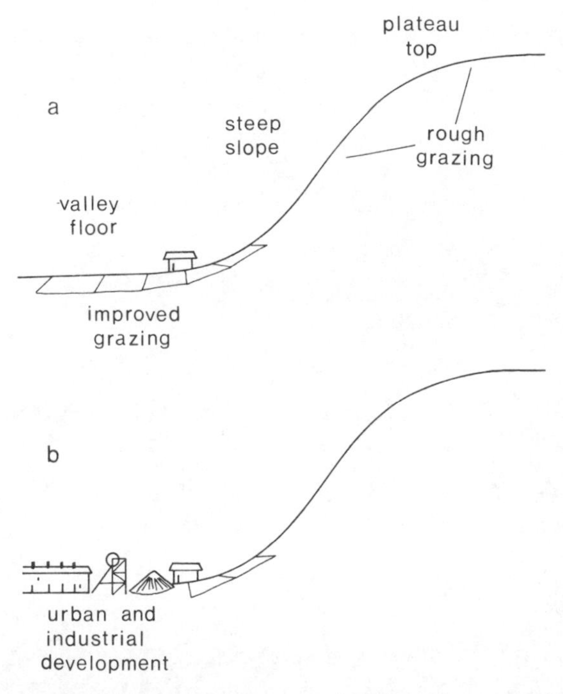


*Fig. 9.11* The structure of a valley farm in (*a*) rural setting and (*b*) urban setting

*Fig. 9.12* Windthrow of sitka spruce in the Afan Valley. Waterlogging of the soil on the plateau top has prevented deep root development.

the valley, probably due to the leaching out of mineral nutrients from the soil (Fig. 9.13).

*Fig. 9.13* Poor tree growth on steep slopes at the head of the Afan Valley, probably associated with the leaching of nutrients from the soil

On the farming side, it has become apparent in recent years that the hill tops, far from being unimprovable, can be developed to serve some of the functions originally served by the lost fields on the lower slopes. For example, at Gelli farm in the Afan Valley, moorland areas on the plateau top, at an elevation of 550 m and with an annual rainfall of 2280 mm, have been converted successfully into good grazing land by a careful programme of fertiliser application and controlled grazing in newly enclosed fields.

This suggests that following industrial development in the valley, there was scope for a planned apportionment of land between forestry and farm enterprises in which the farms were balanced units with proper perimeter fences. In these circumstances the animal-straying problem would never have arisen. Instead the valley has experienced an incomplete and haphazard forestry expansion which has added to the disruption of farm structures; it is this disruption that has caused the straying problem. The situation reflects the general absence in Britain of any proper planning machinery to regulate the interplay between forest and farm development and in these circumstances there has been little incentive to make the necessary comparisons of the ecological requirements of the two enterprises.

### The recreational use of the river

Included in the recreational plans for the valley's Country Park were proposals for river bathing facilities, riverside picnic areas, and the construction of a series of bathing and paddling pools filled from the river. Although on superficial examination the river water looked clean, an obvious reason for questioning the wisdom of these proposals was the presence of storm water overflow pipes from the trunk sewer at a number of points upstream of the Park. A study was therefore, made of the degree of sewage pollution in the river to provide a basis for judging whether this particular recreational development was advisable. In 1973 faecal coliform counts indicated a high level of faecal contamination at many of the stations in the main river, including the Country Park site (Table 9.1; Fig. 9.14).

The type of sewage system operating in the valley is designed to overflow into the river when it is overloaded, which is usually during rain storms when there is a large influx of surface water into the sewers. The discharges take place through storm water overflows of the type shown in Fig. 9.15. When a system of this kind is working properly, although it discharges faecal material into the river, it should do so only in storm conditions when both river and sewer are full and conditions are most favourable for wastes to be

*Table 9.1* Results of faecal coliform counts (numbers/100 ml) at different stations in the River Afan system, January–July 1973. Station 5 is the Country Park (based on data collected by G. H. Tarbutt, University College, Cardiff)

| *Station* | *Maximum* | *Minimum* | *Median* |
|---|---|---|---|
| 1 | 54 000 | 2 400 | 24 000 |
| 2 | 24 000 | 790 | 11 000 |
| 3 | 24 000 | 3 300 | 17 000 |
| 4 | 3 400 | 20 | 850 |
| 5 | 35 000 | 20 | 230 |
| 6 | 54 000 | 170 | 1 100 |
| 7 | 3 500 | 130 | 330 |
| 8 | <20 | <20 | <20 |
| 9 | 16 000 | 1 400 | 3 500 |
| 10 | <20 | <20 | <20 |
| 11 | 3 500 | 170 | 700 |
| 12 | 5 400 | 1 700 | 3 500 |
| 13 | <20 | <20 | <20 |

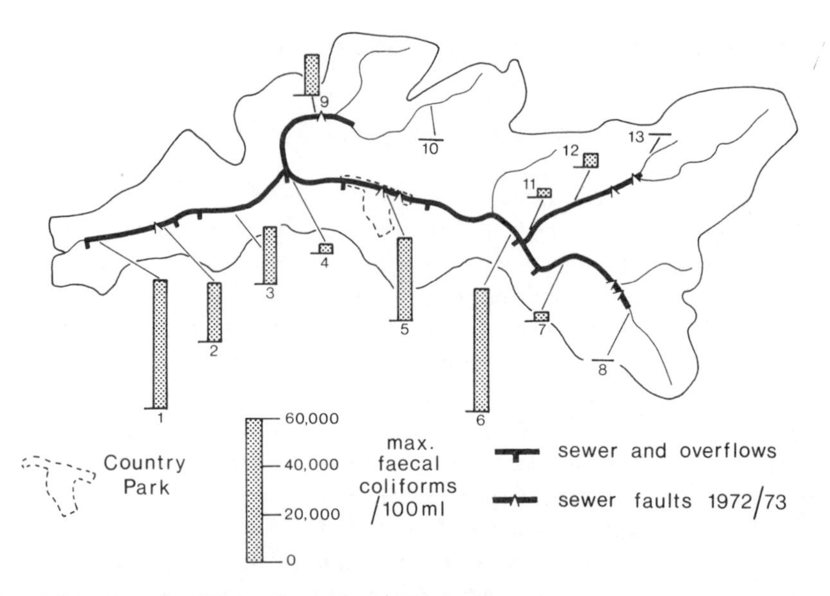


*Fig. 9.14* Faecal coliform counts in the River Afan system

*Fig. 9.15* A storm water overflow

diluted and carried away. In the Afan valley, however, high counts were recorded at many stations along the river even in low flow conditions when the storm water overflows were not discharging.

The reason for this became apparent when the sewage system was examined in detail along its length. The main sewer was fractured in some places, partially blocked in others. There were also some lateral sewers which had never been connected to the main trunk sewer and which discharged directly into the river. During 1972 and 1973 four fractures, two blockages and two unconnected pipes were located (Fig. 9.14). Ironically when the toilet facilities for the Country Park Information Centre itself were constructed, the sewer to which they were linked was found to be broken and leaking onto the river bed alongside the proposed picnic site and bathing area. As has been described earlier (p. 45) the appropriate standard of cleanliness recommended for recreational waters is a matter of some debate. However, virtually all authorities would reject a site, such as the proposed swimming area on the River Afan, which had faecal coliform counts up to 35 000/100 ml. In fact, a count of only a 100 faecal coliforms would make any public bathing pool doubtfully acceptable.

The faecal coliform/faecal streptococci ratios at some stations on the river indicated an important contribution from the faeces of animals other than man. Sheep are the most likely culprits and as many town pavements are often covered with sheep droppings, bacteria from this source must constantly be entering the river through the storm water overflows. The contamination of the river by drainage from normal sheep pastures is probably negligible compared with that from paved areas. It is, however, of little consolation that some proportion of the faecal coliform count in the river is attributable to the faeces of sheep rather than those of human origin. Although sheep do not carry serious pathogens such as typhoid organisms they do carry organisms which can produce minor enteric infections. Children already bathe in some contaminated sections of the river and in spite of the difficulties of interpreting coliform counts in health terms, it is probably significant that outbreaks of minor enteric disorders frequently occur during the summer bathing period.

On the basis of these findings the river bathing project in the Country Park was abandoned. Instead a simple paddling pool was created on a clean side-stream and the development of the Park was modified to give increased emphasis to the forest areas away from the river. Since the survey in 1973, some of the defects in the sewage system have been remedied by the Local Authority and there has been a considerable improvement in faecal coliform counts in the Country Park, During 1974 the maximum value was 2100/100 ml, the minimum 20 and the median 210. This is within the range which many authorities would find acceptable.

One must draw the conclusion that no recreational bathing projects should be seriously contemplated in areas with old-fashioned storm-water overflow systems, until proper bacteriological surveys have been made and, where necessary, remedial measures have been carried out.

There is a further objection to the recreational use of rivers in these valleys which arises from the disposal of domestic rubbish. Although there is an organised system of rubbish collection, it is not uncommon for householders and tradesmen living close to the river to throw rubbish onto the river banks. These rubbish tips attract rats which can frequently be seen even during daylight hours. Rats can carry the bacterium *Leptospira icterohaemorrhagiae* and excrete it in their urine. In such situations drainage from the tips will carry the organisms into the river. In man these bacteria cause a severe, often fatal, infection known as Weil's disease (p. 54). The infection is contracted when organisms from contaminated water enter the human body via cuts and abrasions. There have been a number of cases in South Wales, including one in the Afan Valley attributable to this route of infection. The simple solution is to prevent indiscriminate tipping of rubbish along river banks.

### *Angling*

The River Afan once supported thriving populations of migratory fish. In 1706 for example 37 salmon and 559 migratory trout were caught in the lower part of the river. Brown trout must also have been common at this time. Industrial development had a catastrophic effect on the river fishery. It was not until the coal industry had declined in the 1950's that the fishery had recovered sufficiently to stimulate the formation of the Afan Valley Angling Club.

Onc of the adverse effects of the coal industry arose as a result of coal washing and the anglers' complaints about coal dust in the river continued up until 1970 when the last coal washery was closed. It is well known from observations in other areas that coal particles can affect trout in a number of ways. In the first place, particles settling on the river bed can make it uninhabitable for many of the invertebrate animals on which the fish feed. The particles also clog the coarse gravels used by the fish for spawning so that they no longer provide the conditions of free water circulation and good oxygen supply required by the developing eggs. Additionally, high concentrations of particulate matter (in excess of about 80 mg/l) affect the fish directly by causing damage to the gills.

A survey of the Afan system showed that this kind of contamination had virtually ceased with the closure of the coal washeries; some fine material still washes from river-side spoil tips in heavy rain, but the amounts are small. However this may be only a temporary reprieve. Derelict spoil tips can contain as much as 16 per cent coal and a number of tips elsewhere in the coal field are being reworked to provide small coal for power stations. The reworking operation involves a washing process from which fine coal particles are settled out in lagoons. On a number of occasions water escaping from lagoons has contaminated a river with coal particles. A number of the tips in the Afan valley are suitable for reworking on the basis of the coal they contain. There have also been proposals for reopening some of the Afan pits because of the present fuel shortage. If any of this comes to pass, the associated coal-washing operations will need to be strictly controlled to protect angling interests.

*Mine water*

The coal industry has recently affected the river in a different and more unexpected way. When the drift mines were being worked, water which percolated into the workings from the surface was removed using pumps. The water discharged into the river was relatively clean. Now that the drift mines are closed, pumping no longer takes place and the water entering the mine builds up slowly underground before overflowing into the river. During this build-up the water picks up a heavy load of dissolved iron from strata inside the mine and when it emerges at the surface the iron is deposited as a bright orange sludge of ferric hydroxide [$Fe(OH)_3$] on the stream bed (Fig. 9.16). The mechanisms involved in the process have been studied in coal-mining areas both in Great Britain and in North America. It is known that the iron dissolved by the water inside the mine is derived from iron pyrites ($FeS_2$) and siderite ($FeCO_3$) and that the acid conditions required for this process are created when sulphur-bearing compounds in the mine are converted to sulphuric acid ($H_2SO_4$) by chemical and bacterial action underground.

The precipitation of the ferric hydroxide which occurs when the water emerges at the surface is brought about, partly by a simple oxidation process as the water becomes aerated and partly by the action of bacteria. When the deposit on the stream bed is examined microscopically it is seen to consist of a mixture of fine particles of hydroxide and bacterial filaments. These filaments are responsible for the slimy consistency of the growth.

In the Afan valley there are a number of sections of the river where the bed has an uninterrupted coating of orange sludge, formed by these processes. Altogether over 10 km of the river system is seriously affected. The most spectacular discharge of mine water is from the old Garth Tonmawr Colliery (Fig. 9.17). The water which escapes from the bricked-up level is very acid (pH 3·5) and has a high dissolved iron content, on average 135 mg/l. The

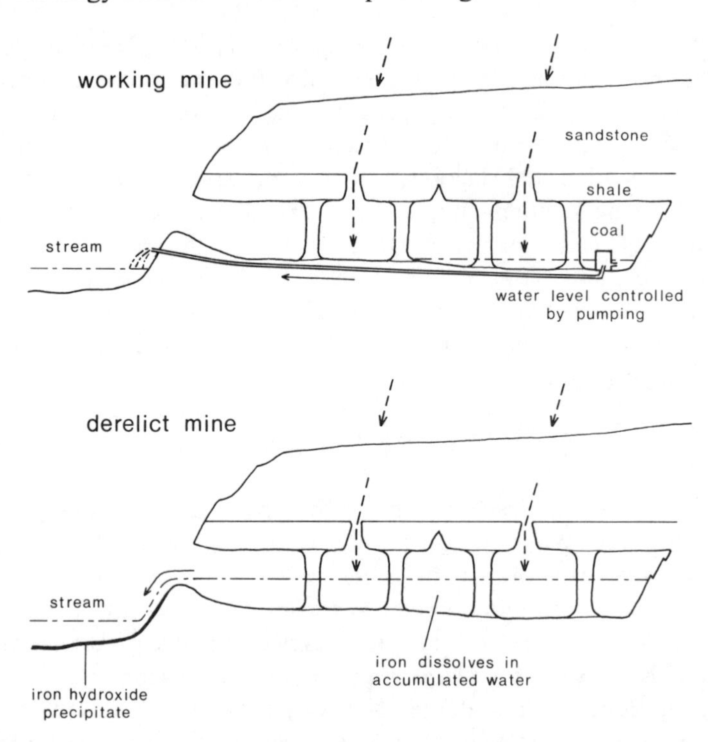


*Fig. 9.16* The formation and discharge of ferruginous mine water from a derelict drift mine compared with the situation in a working mine

discharge is the main cause of the ferruginous deposit which affects the River Pelena all the way down to its confluence with the River Afan (Fig. 9.17). None of the zones affected by the deposit supports a permanent fish population. The ferruginous material probably affects fish in the same way as coal particles, by interfering with their respiration, by contaminating spawning areas and by blanketing out food organisms. This latter effect is well shown in Fig. 9.17 where the mean density of invertebrate animals at a series of stations is compared with the concentration of dissolved iron. In general the greater the iron concentration the more impoverished is the fauna. At concentrations in excess of about 2 mg/l only a few resistant groups persist (the larvae of stone flies and chironomids) and even these groups virtually disappear above about 10 mg/l.

The critical iron concentration in rivers affected by mine water seems to be determined in part by the acidity of the water. Thus in the nearby River Kenfig system, which is also contaminated by mine water but where the water is less acid (pH 7), invertebrate and fish populations are able to live in higher iron concentrations, in fact up to 5 mg/l. This means that, if techni-

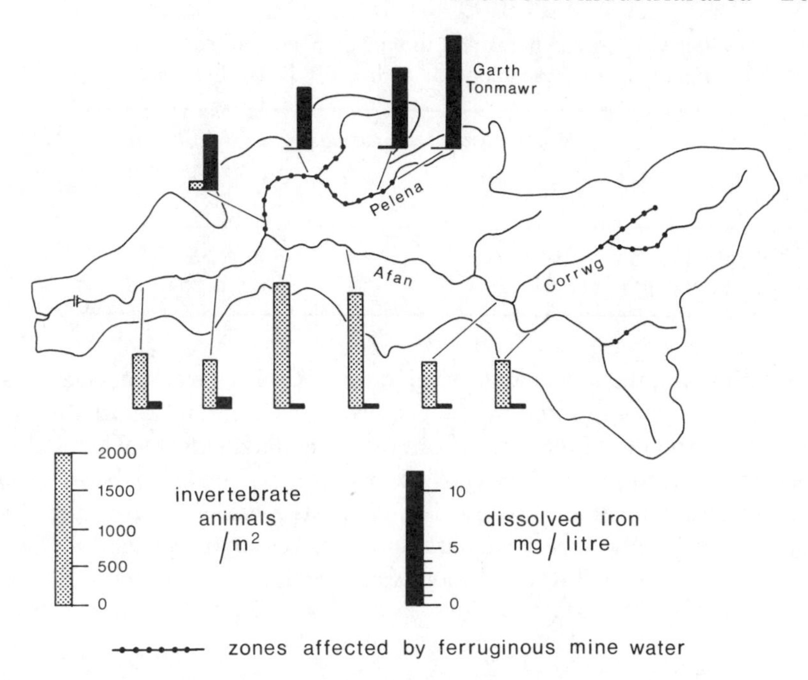


*Fig. 9.17* The incidence of ferruginous mine water pollution in the River Afan system. There is an inverse relationship between the concentration of dissolved iron and the density of invertebrate animal populations (based on data collected by G. L. Wills, University College, Cardiff)

ques are developed to treat mine water discharges, some target levels can be set for the iron and acidity levels which must be reached before these river sections can be reclaimed for angling purposes.

The possible treatment methods are of some interest. One suggested remedy is to sink a new shaft into the old workings and to pump water to the surface by means of a submersible pump. After cleaning the backlog of polluted water in the mine it is envisaged that the water would run clean again, as it would not have had time either to become acid or to dissolve appreciable quantities of iron. In essence this approach aims to revert to the situation when the mine was working. Another alternative is to attempt to precipitate the iron at the surface by treating it with lime. The main problem with this method would be the disposal of the sludge produced.

## *Impediments to fish migration*

In addition to the impact of coal mining, port and industrial developments at the river mouth have also had a deleterious effect on angling in the river by

*Table 9.2* A comparison of the recommended dimensions for fish passes and those adopted for the Port Talbot fish pass

| | *Recommended dimensions* *Metres* | *Port Talbot fish pass dimensions* *Metres* |
|---|---|---|
| Pool lengths | 3·0–6·0 | 0·9–2·8 |
| Pool depths | 0·9–1·2 | 0·7 maximum |

interfering with the movements of migratory fish. The direct passage of fish up the river was blocked when arrangements were made to divert a substantial amount of the river's flow, first to maintain the level in the dock basin and later to supply the steelworks with water. The fish pass which has recently been built on the diversion weir departs considerably from the ideal dimensions, having pools which are both too short and too shallow (Table 9.2). However, even if its dimensions were ideal, so much water is diverted

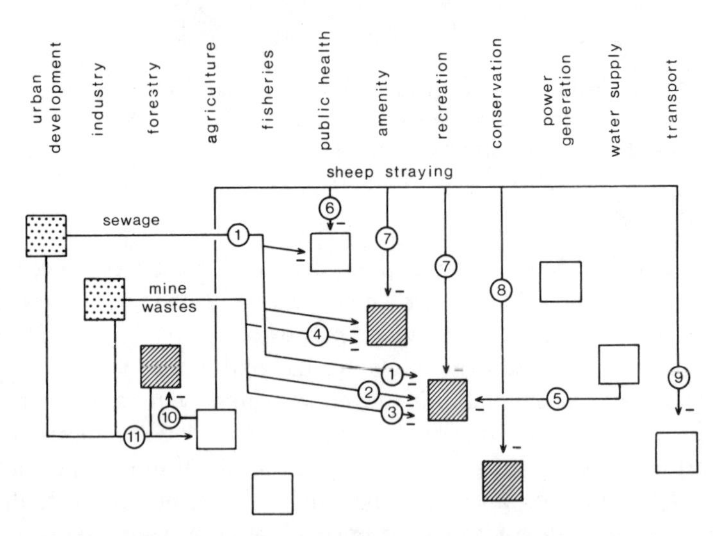


*Fig. 9.18* Ecological interactions in the Afan Valley. Striped blocks show present planning strategy, dotted blocks the earlier one

*1.* Sewage contamination of potential bathing waters
*2.* Effect of coal washing on angling
*3.* Effect of ferruginous mine water on angling
*4.* Disfigurement of landscape by spoil tips
*5.* Interference with fish migration from industrial water abstraction
*6.* Involvement of sheep in echinococcosis
*7.* Sheep damage to gardens and landscaped areas on tips
*8.* Sheep damage to oak woodlands
*9.* Sheep interference with road traffic
*10.* Sheep damage to forestry plantations
*11.* Disruption of farm units by urban, industrial and forestry development

for industrial purposes upstream that the volume of water remaining would rarely be sufficient to service the pass.

### The ecological commentary

The study showed that until remedial measures to solve the problems of persistent river pollution and extensive dereliction are undertaken throughout the valley, local schemes for amenity improvements and recreational developments are likely to be impeded (Fig. 9.18, *1–4*). This is a familiar pattern in many areas where once flourishing industries have declined. Remedial action is often hampered by high costs and disputes concerning areas of responsibility. This points to the need with modern industries to cater for environmental protection from the outset.

In the Afan valley the rehabilitation schemes are also seriously hampered by the numerous side effects of sheep-straying (Fig. 9.18, *6–10*). On closer examination this problem too was found to have its origin in the region's sudden urban and industrial growth. The disruption of the local rural economy set in motion an incomplete and haphazard transfer of land to forestry, with all its attendant side effects. Unfortunately most planning systems are still incapable of controlling rural land-use or regulating the wider impacts of urban and industrial development.

# 10 A tropical development area

As a contrast the fourth study was located in a developing country in the tropics. A recurrent pattern through the tropics is for urban growth to outstrip the capacity of service systems, particularly those concerned with water supply and waste disposal. The hazards of poor sanitation are thereby added to the traditional threats from tropical diseases. Many of the expanding towns also have food supply problems because the surrounding country has been developed for the production of cash crops and yields little food for local consumption. The problem tends to be aggravated by the migration of farmers and their families into the towns and by inadequate arrangements for storing food and transporting it from rural areas.

Southern Nigeria has its share of all these problems. For example the city of Ibadan (pop. 650 000) is sited in the food deficit area created by the western cocoa belt and has to rely on food brought in from the north and east (Fig. 10.1) (Agboola, 1971). With further expansion, Benin (pop. 130 000) is likely to face similar problems because the surrounding region has been largely committed to timber and rubber production. The natural vegetation type over much of southern Nigeria is high forest (Fig. 10.1*b*) and considerable areas have so far escaped the attention of timber extractors, farmers and hunters, or have had time to recover from such interference. However the future of these remnants is in doubt because western concepts of conservation find no general acceptance amongst the public at large and have only recently been recognised in official circles.

The area around Benin was chosen for a detailed study. Benin is an ancient city which is now expanding rapidly and developing its first industrial estates (Figs. 10.1, 10.2, 10.3). The environmental problems of the city and its region serve to illustrate how ecology can contribute to planning in a tropical situation.

## Urban growth and public health

### *Water supply*

Many of the growing towns in Southern Nigeria are being provided with piped-water supplies, thus removing the need to collect water from surface water courses and wells. Whilst most of the local people agree that the new water supplies are convenient, not all are convinced that they are pure. Newspapers frequently carry advertisements which urge the housewife to disinfect the family's drinking water and many Nigerians take the precaution of boiling and filtering all the drinking water used in their households. One microbiologist writing recently about the water supply in his University felt it appropriate to ask the question 'Drinking water or sewage, is there a

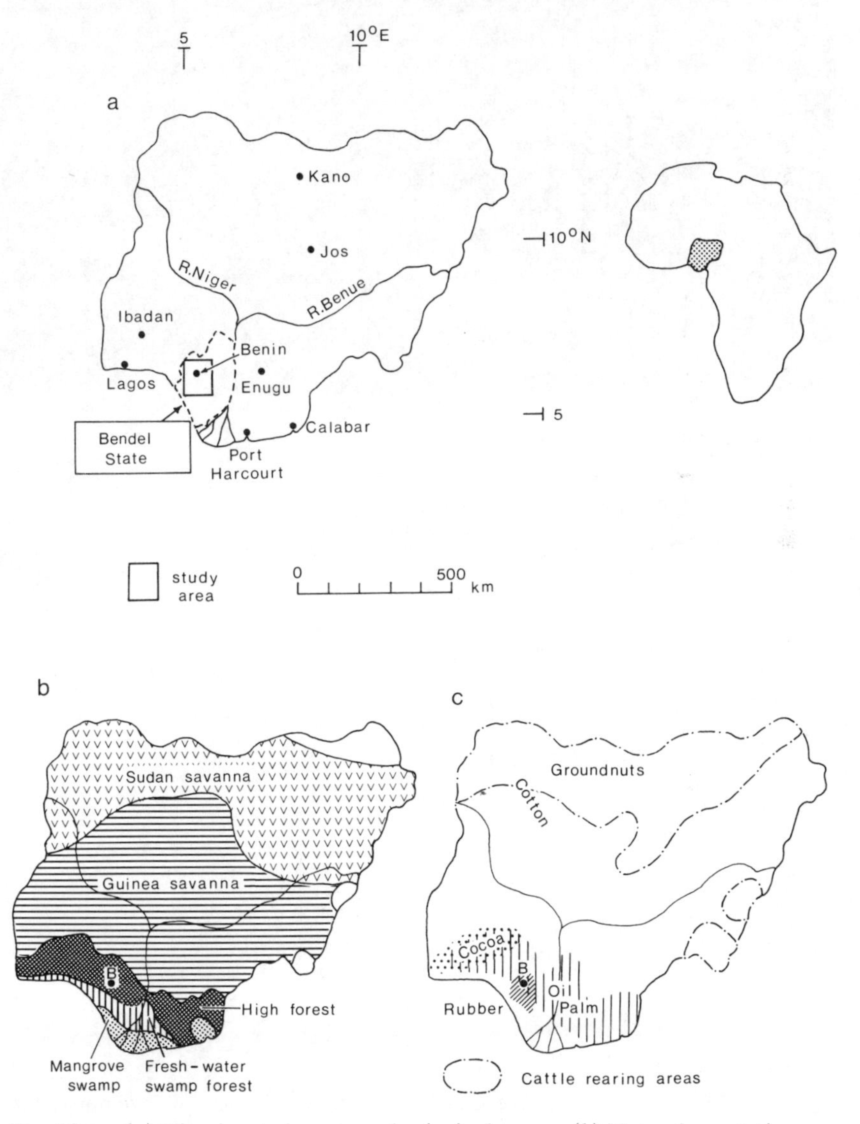


*Fig. 10.1* (*a*) Nigeria, study area and principal towns. (*b*) Natural vegetation zones. (*c*) Cash crops

*Fig. 10.2* Benin City

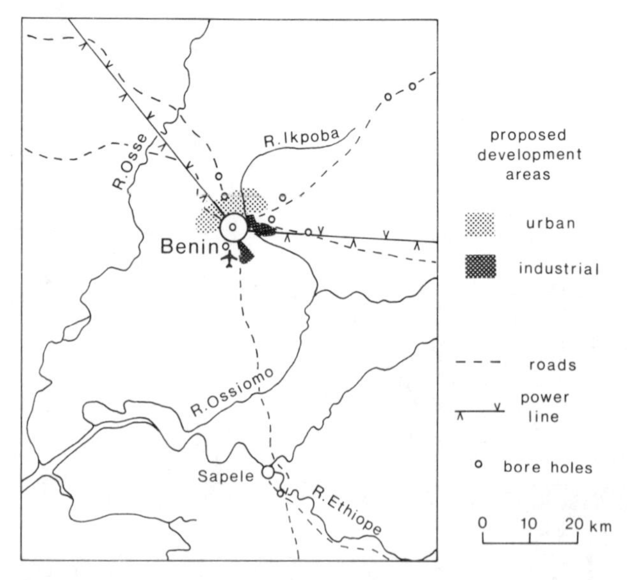


*Fig. 10.3* Service systems and proposals for urban and industrial development in the Benin region

difference?' (Damann, 1973). In that instance it was shown, using routine bacteriological tests, that the drinking water was in fact frequently contaminated by sewage.

The water for Benin is derived largely from boreholes which tap underground sources some 60 m beneath the city. After passing through a treatment plant the water is piped to street taps (Fig. 10.4*a*) and, in some parts of the city, to taps in private houses. Although at first sight these arrangements seemed to allow little opportunity for serious contamination, bacteriological tests showed that somewhere faecal material was entering the system. Presumptive coliform tests and plate counts (using eosin-methylene blue agar) were carried out on water samples from a number of street and house taps. Some of the samples gave presumptive coliform counts in excess of 1800/100 ml of water and on many of the plates more than 20 colonies of *Escherichia coli* type 1 developed from a 1-ml inoculum indicating that the high numbers from the total coliform test were predominantly due to *E. coli* 1. To put these results into perspective, the World Health Organisation recommendations for drinking water standards suggest that presumptive (total) coliform counts should not exceed 10/100 ml of water and that at all times drinking water should be free of *E. coli* 1 (World Health Organisation, 1971).

There was no doubt that serious, intermittent contamination of the water supply was occurring. In searching for its source a visit was made to the pumping station at Ogba. There it was discovered that water drawn from bore holes was supplemented from time to time by water from a spring source (Fig. 10.4*c*). Whilst spring sources are usually expected to yield bacteriologically clean water, this was not the case with the Ogba spring and its water gave presumptive coliform counts in excess of 1800/100 ml. Further investigation showed that the spring originated from a perched water table lying at a relatively shallow depth (about 10 m) underneath villages to the west of the city. The village pit latrines were thus likely sources of contamination of the spring (Fig. 10.5). The pumping station had facilities for treating water with gaseous chlorine but it was apparent that the volume of water being handled often exceeded the capacity of the treatment plant. Also at times of heavy demand the treatment plant was often bypassed altogether.

A further complication was revealed when water from some house taps was tested. It was found that taps in the same house and supplied by the same water main could yield water of different qualities. In one house, the bathroom tap produced clean water but the kitchen tap, contaminated water. The explanation here seemed to be that within the house compound, seepage from the soakaway (Fig. 10.4*b*) was entering the water pipe either through cracks or through poor pipe joints. Elsewhere in the city, pit latrines

a

b

*Fig. 10.4* (*a*) A typical street tap in Benin City; (*b*) A soakaway of the kind used for waste disposal in some parts of the city; (*c*) The Ogba pumping station, where borehole water is mixed with water from a local stream before entering the supply system, the top of the borehole can be seen on the left

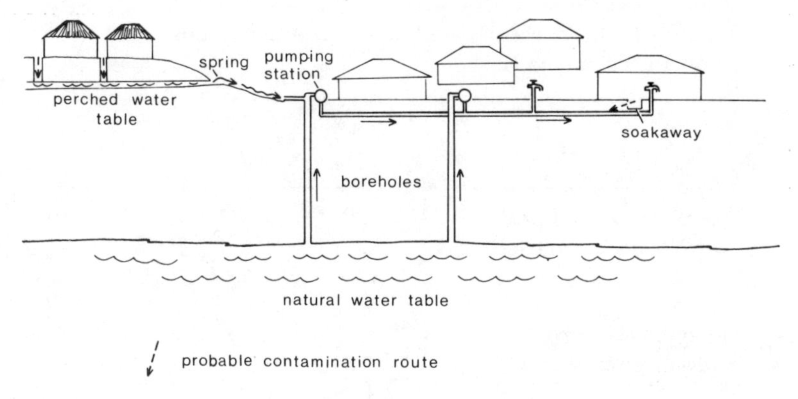


*Fig. 10.5* Some features of the Benin water-supply system

are likely sources of contamination. Normally the positive water pressure in the pipe would ensure that any seepage would be outwards from the pipe. However where the water supply is intermittent, as it is in Benin because of electrical power failures, polluting material can pass in the opposite direction and contaminate the supply.

Thus the provision of a piped supply is not in itself a guarantee of pure water, particularly in situations where the risk of contamination is high. It is clearly necessary to have a treatment plant which keeps pace with increasing water demands and to institute a routine bacteriological monitoring programme to pick up faults in the system as soon as they occur.

## *Waste disposal*

In addition to enteric disorders attributable to water-borne pathogens, many town dwellers in Southern Nigeria also suffer from roundworm infections. This was shown in a study of six towns carried out by Hinz (1967) and the same picture emerged from an examination of the records of the Specialist Hospital in Benin (Table 10.1). Roundworm parasites enter the body either as eggs taken in with food (*Ascaris* and *Trichuris*), or as larvae which bore through the skin (*Strongyloides* and the hookworms). The parasite problem is certainly one to be taken seriously; the hookworms can cause anaemia as the result of intestinal bleeding and the adults of all the species interfere with digestive processes in one way or another. In addition the migration of larval hookworms and *Ascaris* can cause lung damage (World Health Organisation, 1964).

As infective eggs and larvae both originate from the faeces of infected persons it might be thought that the incidence of the parasites reflects some basic deficiency in waste disposal systems. In fact this is not the main cause of the problem; the pit latrines, bucket latrines and flush toilets in use in the

*Table 10.1* Incidence of parasitic roundworm infections in Southern Nigerian towns

| *Parasite* | *A** % | *B*† % |
|---|---|---|
| *Ancylostoma duodenale* or *Necator americanus* (hookworms) | 20 | 23·6–40·7 |
| *Strongyloides stercoralis* (dwarf threadworm) | 3 | Not assessed |
| *Ascaris lumbricoides* (large intestinal roundworm) | 19 | 4·5–36·9 |
| *Trichuris trichiura* (whipworm) | 9 | 3·3–26·7 |

*Column A: infections in 300 patients attending the Specialist Hospital at Benin, March–April, 1974

†Column B: range of infection levels in six towns studied by Hinz (1967)

towns are perfectly adequate for their purpose. The main difficulty arises from the fact that some members of the population do not use them and defaecate at random on urban waste land and in farm plots during the course of their daily activities. Many children are allowed to use the house compound for this purpose. This creates favourable conditions for the spread of parasites because viable parasite eggs can be transferred to food by soiled hands and hookworm larvae have the opportunity to bore through the skin of unshod feet. Seen in this light the alleviation of the problem seems to lie more in the field of public health education rather than in town planning and sanitary engineering.

## *Domestic refuse*

In temperate countries the main problems associated with refuse disposal come from the scavengers such as gulls and rats which are attracted by the edible material contained in tips (p. 49). Scavengers are not necessarily a problem in tropical towns because edible scraps are less readily discarded. In Benin most of this material is fed to goats and hens and very little gets as far as the tip.

There are however some specifically tropical problems related to the role of tips as mosquito-breeding sites. Small accumulations of water in derelict car bodies and in tins on refuse tips serve as important breeding sites for the yellow fever mosquito *Aedes aegypti* (p. 49) and there is a general case for reducing the availability of these situations whenever possible. In Benin however, the prospect of controlling *Aedes* effectively is remote because the many trees in the town (Fig. 10.2, p. 210) provide innumerable water-filled tree-holes.

*Industrial development*

Most of the well-established industries in the Benin area are concerned with processing local natural products such as timber and rubber and their environmental impact has been negligible. The only waste from a crêpe-rubber factory, for example, is waste water containing grit and small rubber fragments; both materials are quickly dispersed by discharging the water to a river. With more recent developments the emphasis has been on consumer goods and this has led to the establishment within the city of a vehicle assembly plant, bottling plants and breweries, and factories to produce carpets and floor tiles. Possible future developments include glass, leather, metal and paper industries.

Two sectors of the city perimeter have been designated for industrial expansion, one to the south along the road to Sapele and the other to the east on higher ground adjacent to the banks of the River Ikpoba (Fig. 10.3). This latter site has been favoured because of the possibility of discharging industrial wastes into the river. Although these industrial developments are on a modest scale, many local observers are concerned about environmental damage from industrial pollutants. This unease has led to the establishment for the region of an Environmental Protection Agency based on the American pattern. The potential conflicts are well illustrated by the River Ikpoba site where polluting discharges would seriously interfere with other uses of the river (Fig. 10.6). These include fishing, cattle-watering, clothes-washing and bathing.

Pollution problems associated with different industries have been widely studied in Europe and North America and this information is readily available (Hynes, 1960; Klein, 1962). There is, however, the drawback that the information relates specifically to temperate situations and may have to be interpreted in a somewhat different light for the tropics. For example, brewery effluents contain substantial amounts of organic matter and it is known for temperate rivers that the decomposition of this material can be harmful to fish because it depletes the water of oxygen. Whether these effects are likely to be more serious in warmer tropical rivers is not clear. Equally we have insufficient information on the oxygen needs of tropical river fish. Some species at least may be tolerant of deoxygenated conditions

*Fig. 10.6* The River Ikpoba at the edge of the city is used for fishing, watering cattle, bathing and clothes-washing

because they have encountered them earlier in their evolutionary history in swamp habitats.

In the tropics, environmental changes must always be examined for their possible effect on disease patterns. Organically-rich brewery effluents may certainly be of significance in this context because of the ecological attributes of the mosquito *Culex pipiens fatigans*. It will be recalled that this species has a special preference for organically-rich water as a breeding site and is a carrier of the filarid worm (*Wuchereria*) which causes elephantiasis. Benin has two large breweries on the Ikpoba industrial estate which discharge their wastes down a single concrete channel into an area of swamp forest. From there the effluent finds its way to the river. The swamp forest pools have been so enriched by this process that they now form a major breeding focus for *C.p. fatigans* and many hundreds of larvae and pupae can be caught with a single sweep of a pond net (Fig. 10.7). In the presence of *Wuchereria* the site could form a major health hazard close to the city. Fortunately the parasite does not seem to have spread to the Benin area to date and one hopes that, before it does, steps will have been taken to improve the standards of effluent treatment and disposal.

*Fig. 10.7* Swamp-forest pools contaminated with brewery effluents form a major breeding site for the mosquito *Culex pipiens fatigans* at the edge of the city. The channel carrying the effluent from the brewery is seen in the background.

## Cash crops and food supplies

The Benin region has all the attributes of a potential food shortage area. Large sectors of the countryside surrounding the city are devoted to timber, rubber and oil palm production (Fig. 10.8). All these are cash crops, orientated to export markets. The total area remaining for food-producing farms is limited. The farm holdings themselves are small and the produce from each generally does little more than serve the needs of the farmer and his immediate family. Table 10.2 gives some statistics for small farms typical of this region associated with the village of Akumazi, east of Benin. Consequently most of the staple foods consumed in the city, such as cassava (*Manihot utilissima*), yam (*Dioscorea spp.*) and rice (*Oryza sativa*) have to be brought in from further afield, principally from the Guinea savanna zone to the north-east where food-crop production is the principal land use.

Livestock production on Benin farms is negligible and the meat requirements of the city are met from beef cattle brought down by Fulani tribesmen from Northern Nigeria and from 'bushmeat'. Almost any animal caught in

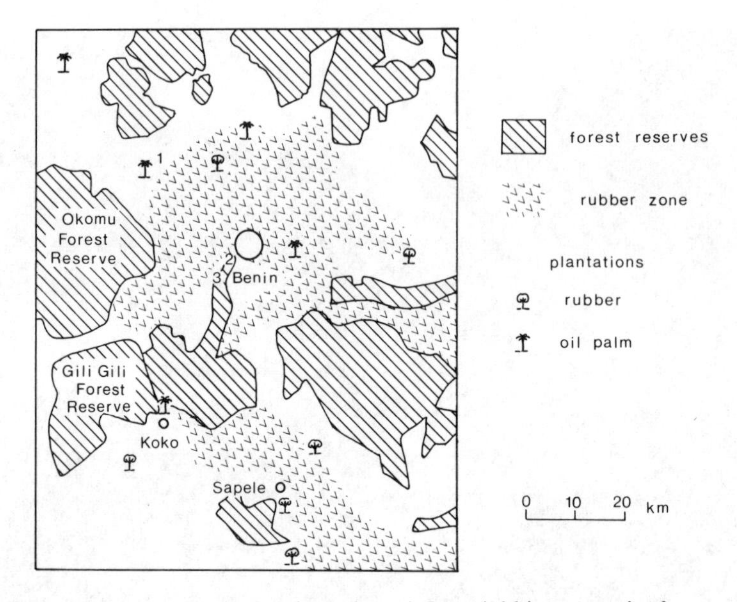


*Fig. 10.8* Cash-crops in the Benin region: *1.* Iguoriakhi community farm; *2.* Ogba fish ponds; *3.* Ogba pig and poultry farm

*Table 10.2* Farm statistics for the village of Akumazi (after Upton, 1967)

| *Field crops (% of arable land under each crop)* | | *Tree crops (Average number of trees/farmer)* | | *Livestock (Average number of animals/farmer)* | |
|---|---|---|---|---|---|
| *Crop* | % | | | | |
| Cassava | 25·7 | Oil palms | 72·0 | Cattle | 0·3 |
| Yam | 23·6 | Bananas | 35·0 | Goats | 1·0 |
| Cocoyam | 12·1 | Rubber | 4·5 | Poultry | 5·5 |
| Rice | 8·7 | | | | |
| Maize | 7·1 | | | | |
| Peas and beans | 0.7 | | | | |
| Other (e.g. peppers, gourds) | 22·1 | | | | |

| | |
|---|---|
| Number of years of consecutive cropping (modal value) | 2 |
| Number of years fallow (modal value) | 4 |
| Mean farm area/cultivator | 1·6 hectares |

the forest or on a farm plot comes into this latter category and the list includes mammals, birds, reptiles, molluscs and insects (Ene, 1963; Asibey, 1974). On the approach roads to the city, monkeys, forest antelope (*Cephalophus maxwelli*) and large rodents such as 'the cutting grass' (*Thryonomys swinderianus*) are regularly offered for sale by hunters. Fish caught in the River Osse and the delta area are also available in the markets.

Although at present the city's food requirements can be met, further urban expansion would certainly cause problems. It is with this in mind that agriculturalists have been exploring the possibilities of increasing local food production. Various measures such as community farm projects and farm settlement schemes, have been introduced to rationalise farm structures and to allow greater mechanisation. In addition fish-rearing ponds and intensive pig and poultry production units have been set up at Ogba on the outskirts of the city (Fig. 10.8). Many more such projects will be needed if food production is to keep pace with urban growth.

*Possible side-effects of agricultural and fisheries development*

In most parts of the world where agricultural practices have been intensified, the control of pests and weeds has become an essential part of the operation. Single-species crops are particularly vulnerable. In some instances the control measures themselves have unwanted side-effects on other interests and in advanced countries there has been much concern about adverse effects on species of conservation interest.

So far there is little sign of these difficulties in the forest zone of Southern Nigeria. In the first place none of the major African pest species extends into this area. The desert locust (*Schistocerca gregaria*) and the red locust (*Nomadacris septemfasciata*) do not normally move outside the grassland zone and plagues of the migratory locust (*Locusta migratoria*) are a rare occurrence. The only related species which is spoken of as a pest in the forest region is the variegated grasshopper (*Zonocerus variegatus*). Although it frequents farm plots, it rarely seems to reach its full potential as a pest, possibly because its life cycle is partially out of phase with the crops it attacks. Moreover, it can be easily controlled using minimal applications of safe insecticides.

Africa's most important bird pest, the quelea bird (*Quelea quelea*) which causes serious damage to cereal crops in the savanna areas of West and East Africa, is not important in the forest zone. Various other weaver birds related to the quelea do occur and colonies of the village weaver (*Ploceus cucullatus*) are a familiar sight around Benin. Several hundred birds will congregate noisily in the same tree and suspend their elaborate woven nests from the branches. The village weaver annoys farmers by stripping leaves from oil palm trees to use for nest building and as a result eventually killing the tree. The birds may also damage rice crops but only on a local scale.

The agricultural significance of Southern Nigeria's best known weed, siam weed (*Eupatorium odoratum*), is also arguable (Ivens, 1974). It is an alien species originating in the West Indies and spreading to Nigeria in the 1940's. Although it invades agricultural land during the fallow period it is probably no more difficult to remove than the usual invasive bush plants. It can retard early growth in oil palm and rubber plantations, but does not have this effect on young timber trees and may afford them useful protection in the early stages of growth. In oil palm plantations it is relatively easy to control using herbicides.

Possibly pest and weed problems and their attendant side-effects will become more important in the forest zone with an intensification of agricultural practices, especially where a greater emphasis is being put on single species stands. For this reason the situation needs to be kept under review.

Fish ponds can often make an important contribution to the health of a community by providing a source of protein. However they can also provide suitable habitats for organisms implicated in the transmission of parasitic infections to man and this situation, if left unchecked, can off-set the benefits gained from the additional food resource. Snails involved in the life cycle of the parasite responsible for schistosomiasis (p. 88) are the most important organisms in this category.

The presence or absence of snails in the natural-water bodies of an area can give some indication whether this problem is likely to develop when a new fish pond is established. The Benin region was found to be virtually devoid of fresh-water snails, probably as a result of the low mineral content and high acidity of its natural waters. This fact taken in conjunction with the absence of medical records for schistosomiasis suggests that fish ponds can be constructed with impunity. However, any proposal to increase fish production by the addition of fertilisers to the water would need to be looked at very carefully. Fertilisers could increase the suitability of the habitat for snails and thus introduce the risk of schistosomiasis into a region free of it at present.

## Rural development and biological conservation

The forest habitat which from the air appears to cover much of Southern Nigeria has in fact been extensively modified by human intervention both with regard to its vegetational structure and its animal communities. Richards (1939, 1952) in his classical work on tropical forests came to the conclusion that the Nigerian high forest in its natural form consisted of three tree layers: the uppermost with isolated flat-crowned trees extending to 45 m, a middle layer of narrower-crowned isolated trees up to 37 m and the lowest storey of trees forming a continuous canopy at about 15 m. Beneath this lowest layer of trees would be a layer of shrubs and finally a layer of ground vegetation on the forest floor. Richards used as the basis for his study some forest areas in the Okomu Forest Reserve which even now show many natural features (Fig. 10.9).

*Fig. 10.9* The Okomu Forest Reserve. Although this part of the forest has been modified by road construction and timber extraction, some tall emergent trees still remain projecting above the general level of the canopy

The cultivation of farm plots has had a major impact on the forest (Fig. 10.10). Farmers use a rotational system of bush-fallow in which an area is farmed for a few years and then allowed to return to bush. Each new cultivation phase starts with the bush being cleared by cutting and burning. A typical rotational regime could be three-years cropping followed by six-years fallow, but there are many variations in the timing. The decline in soil fertility during the years of cropping is restored during the ensuing fallow period. A farm plot left uncultivated will eventually revert to forest. However the complete re-establishment cycle takes a long time, of the order of 250 years. During the initial years the plot is invaded by bush growth and a

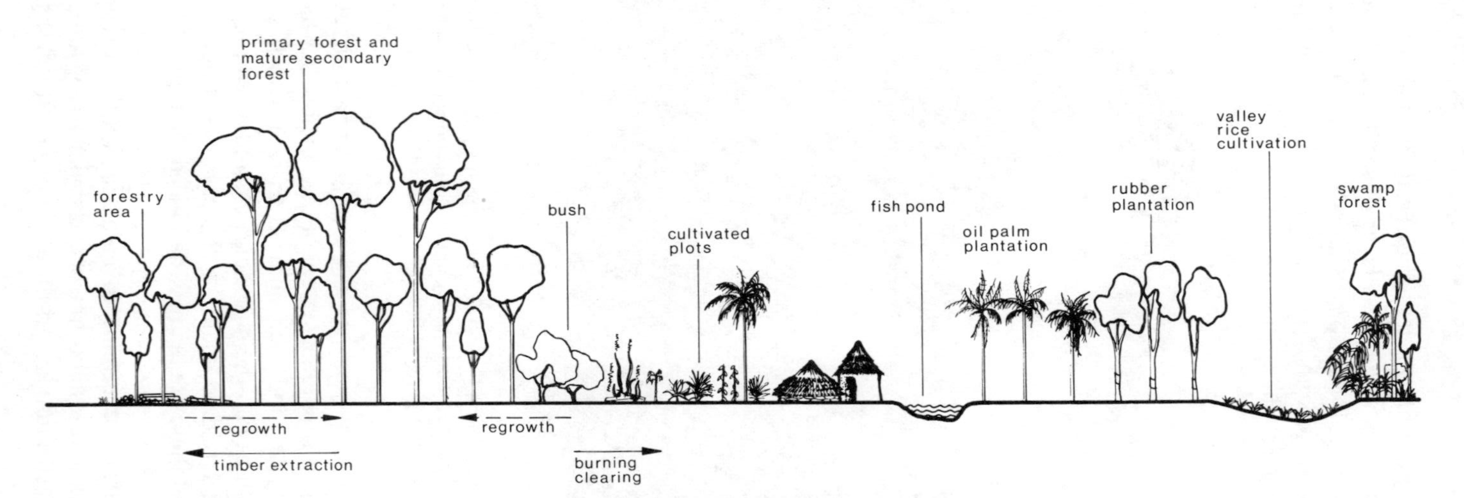


*Fig. 10.10* Diagram illustrating some of the impacts of rural development in the Nigerian forest zone

few tree species which have wind-dispersed seeds, e.g. the umbrella tree (*Musanga cecropioides*) and *Albizzia zygia*. After about 100 years these colonising species will produce a 'secondary forest'. Another 150 years will be needed for the true forest trees, most of them with heavier and less efficiently dispersed seeds, to establish themselves. In heavily settled areas, no farm plot is likely to be left alone for this length of time. In fact, in regions where food is short the fallow period may be reduced beyond the few years necessary to build up soil fertility to a useful level.

The cultivation of rubber and oil palms in plantations has also modified the original structure of the forest. Although indigenous latex-producing trees have been used in the past, modern rubber production is based on the imported Para rubber tree (*Hevea brasiliensis*). Oil-palm plantations similarly represent an artificial habitat, although the oil palm (*Elaeis guineensis*) is at least an indigenous species which occurs in its natural form as isolated trees scattered through the forest.

In theory timber extraction should have only a minor effect on forest structure, because ideally the forester aims to return the timber plot to some semblance of its original structure before the next phase of harvesting. In practice the usual interval between successive timber extractions is too short to allow this to happen. The extraction of large timber also causes widespread disturbance in the forest and produces extensive areas of secondary growth. Additionally the planting of exotic species such as teak (*Tectona grandis*) introduces alien elements into the forest community.

As far as animal populations are concerned, the natural forest supports a rich assemblage of invertebrates, amphibians, reptiles, birds and mammals. Figure 10.11 attempts to portray the main elements of the mammalian fauna. The natural assemblage is more diverse than is popularly imagined, although many species are likely to be seen only well away from areas of human habitation. Many groups are represented by more than one species which often divide the different layers of the canopy between them (Booth, 1960; Napier and Napier, 1967). This is well illustrated by the squirrels (21–23) and the primates (14–17). The original status of some mammal species remains unclear. It is difficult to know for example whether the leopard (*Panthera pardus*) was a true high-forest species or was restricted to the forest edge. Some forest species such as the bongo (*Boocercus euryceros*) are known to occur elsewhere in West Africa but have never been recorded from Nigeria, (Happold, 1973).

Few true forest mammals can persist as permanent residents in farmed areas partly because the physical structure of the habitat is unsuitable and partly because farmers invariably set traps on their land to catch animals for food. The main victims are species such as the cutting grass (*Thryonomys swinderianus*) and the West African ground squirrel (*Xerus erythrops*).

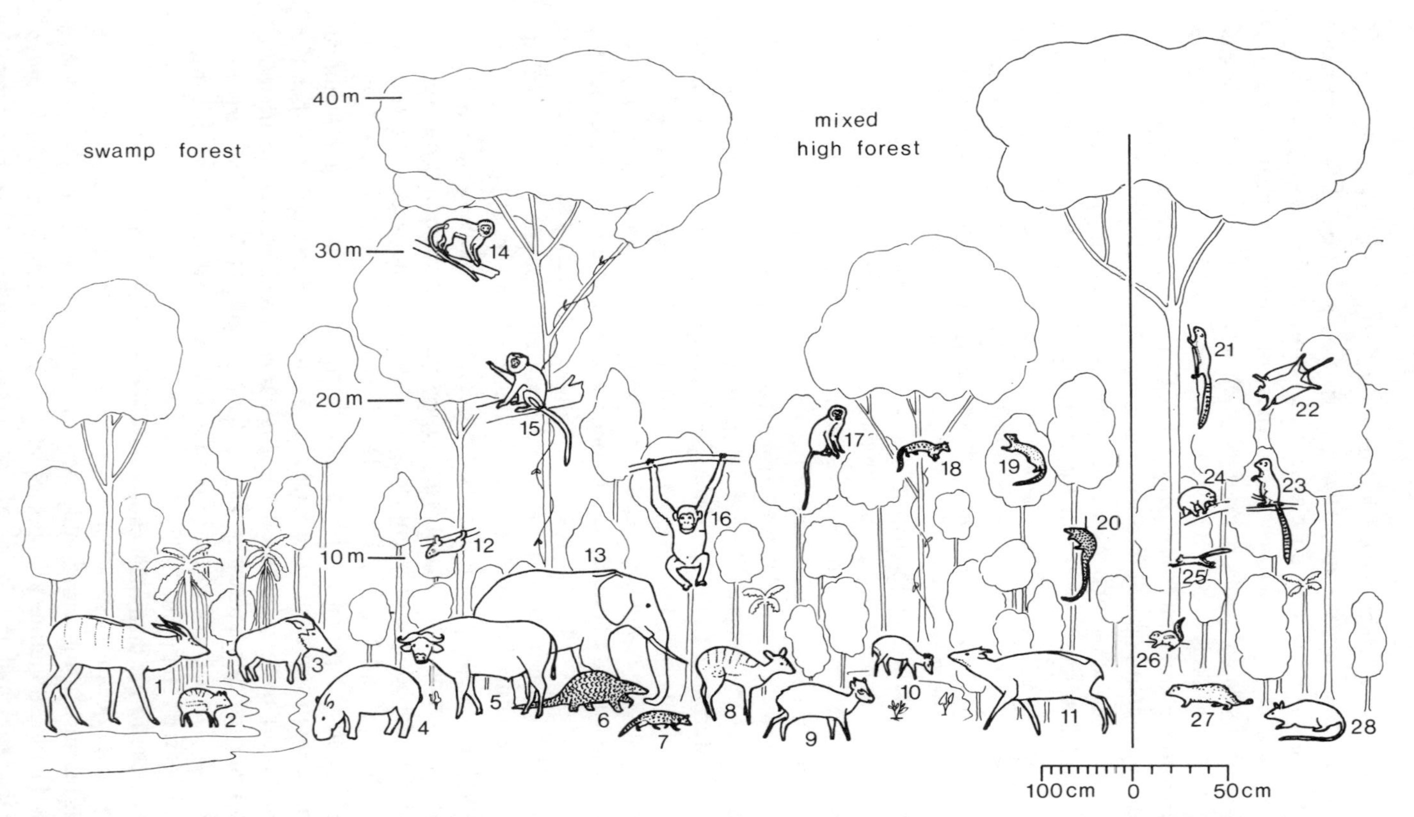


*Fig. 10.11* The natural assemblage of larger mammals in the Nigerian forest zone (different scales have been used for the forest structure and the animals)

1. Sitatunga (*Tragelaphus spekei*)
2. Water chevrotain (*Hyemoschus aquaticus*)
3. Red river hog (*Potamochoerus porcus*)
4. Pigmy hippopotamus (*Choeropsis liberiensis*)
5. African forest buffalo (*Syncerus caffer nanus*)
6. Giant ground pangolin (*Manis gigantea*)
7. African civet (*Viverra civetta*)
8. Bushbuck (*Tragelaphus scriptus*)
9. Black duiker (*Cephalophus niger*)
10. Maxwell's duiker (*Cephalophus maxwelli*)
11. Yellow-backed duiker (*Cephalophus sylvicultor*)
12. Tree hyrax (*Dendrohyrax dorsalis*)
13. Forest elephant (*Loxodonta africana cyclotis*)
14. Mona monkey (*Cercopithecus mona*)
15. Black colobus monkey (*Colobus polykomus*)
16. Chimpanzee (*Pan troglodytes*)
17. Putty-nosed monkey (*Cercopithecus nictitans*)
18. Genet (*Genetta maculata*)
19. Two-spotted palm civet (*Nandinia binotata*)
20. Tree pangolins (*Manis* spp.)
21. Giant forest squirrel (*Protoxerus strangeri*)
22. Flying squirrels (*Anomalurus* spp.)
23. Sun squirrels (*Heliosciurus* spp.)
24. Potto (*Perodicticus potto*)
25. Lesser bush baby (*Galago demidovii*)
26. Side-striped squirrels (*Funisciurus* spp.)
27. Brush-tailed porcupine (*Antherurus africanus*)
28. Pouched rat (*Cricetomys gambianus*)

These are strictly speaking savanna grassland species which have penetrated into the forest zone by exploiting clearings and cultivated land. The few forest species which venture onto farm plots, for example duikers, which sometimes browse on yam foliage, and pouched rats, which gnaw cassava tubers, are also likely to fall prey to the trapper.

Timber production *per se* need do little harm to the forest fauna. For example, managed forest plots close to Benin contain a large proportion of the bird species characteristic of the high forest. However, in the case of mammals, forest roads allow hunters access to previously remote areas and uncontrolled exploitation can threaten the local survival of some species. In various parts of West Africa the palm squirrel (*Epixerus ebii*), the pigmy hippopotamus and a number of monkey species come into this category (Robinson, 1971; Asibey, 1974).

Although it is possible that some of the more resilient species, such as Maxwell's duiker, could be cropped to give a sustainable yield, it is difficult to see how the hunters could be organised to make the necessary periodic adjustments of culling levels which must be an integral part of any controlled exploitation programme. Nor could the co-operation of all the hunters be guaranteed.

One is left with the conclusion that the only effective measure for conserving a sample of the high forest community is to designate reserve areas from which all farming, forestry and hunting activities are excluded. Two sites, the Okomu and Gili-Gili Forest Reserves (Fig. 10.9) have been considered by the Bendel State authorities in this context and a first step has been taken by banning hunting of the bush elephant. However the further implementation of conservation measures will present some difficulties. Timber concessions have been granted for both areas and apart from the ban on elephant hunting there is so far no machinery for controlling shooting and trapping of other species. Any effective conservation programme will need to include the withdrawal of timber concessions from selected areas and the establishment of an efficient warden service to check uncontrolled hunting.

Elsewhere in Africa it has been possible to increase the viability of conservation projects by combining them with tourism. This has been very successful in the savanna areas of East Africa and on a smaller scale in the Yankari Game Reserve in northern Nigeria. Although it would be more difficult in the forest to guarantee visitors a sight of the animals, any scheme for a high forest game reserve would certainly generate international interest and support.

# References

Abdullah, M. I., Royle, L. G. & Morris, A. W. (1972). Heavy metal concentration in coastal waters. *Nature*, **235**, 158–160.

Agboola, S. A. (1971). Nigeria aims at self-sufficiency. *Geogrl. Mag., Lond.* **44**, 43–49.

Alabaster, J. S. (1963). The effect of heated effluents on fish. *Int. J. Air Water. Pollut.* **7**, 541–563.

Alabaster, J. S. (1970). River flow and upstream movement and catch of migratory salmonids. *J. Fish. Biol.* **2**, 1–13.

Alabaster, J. S., Garland, J. H. N., Hart, I. C. & Solbe, J. F. de L. (1972). An approach to the problem of pollution and fisheries. In *Conservation and Productivity of Natural Waters* (Eds. R. W. Edwards & D. J. Garrod), pp. 87–114. *Symp. zool. Soc. Lond.* **29**. Academic Press, London.

Allsop, W. H. L. (1960). The manatee; ecology and use for weed control. *Nature*, **188**, 763.

Allsop, W. H. L. (1961). Putting manatees to work. *New Scient.* **12**, 548–549.

Amdur, M. O. (1971). Aerosols formed by oxidation of sulfur dioxide. *Archs envir. Hlth.* **23**, 459–468.

American Public Health Association (1965). *Standard Methods for the Examination of Water and Wastewater.* 12th ed., New York.

Antonovics, J., Bradshaw, A. D. & Turner, R. G. (1971). Heavy metal tolerance in plants. *Adv. ecol. Res.* **7**, 1–85.

Asibey, E. O. A. (1974). Wildlife as a source of protein in Africa South of the Sahara. *Biol. Conserv.* **6**, 32–39.

Assar, M. (1971). *Guide to Sanitation in Natural Disasters.* World Health Organisation, Geneva.

Attwell, R. I. G. (1970). Some effects of Lake Kariba on the ecology of a floodplain of the Mid-Zambezi Valley of Rhodesia. *Biol. Conserv.* **2**, 189–196.

Bakács, T. (1972). *Urbanisation and Human Health.* Akadémiai Kiadó, Budapest.

Bannikov, A. G. (1961). L'écologie de *Saiga tatarica* L. en Eurasie, sa distribution et son exploitation rationale. *Terre Vie*, **1**, 77–85.

Barber, D. (Ed.) (1970). *Farming and Wildlife, a Study in Compromise.* Royal Society for the Protection of Birds, Sandy, Bedfordshire.

Bayfield, N. G. (1971). A simple method for detecting variations in walker pressure laterally across paths. *J. appl. Ecol.* **8**, 533–535.

Bays, L. R. (1971). Pesticide pollution and the effects on the biota of Chew Valley Lake. *Environ. Pollut.* **1**, 205–234.

Beadle, L. C. (1974). *The Inland Waters of Tropical Africa – an Introduction to Tropical Limnology*. Longman, London.

Beard, R. R. & Wertheim, G. A. (1967). Behavioural impairment associated with small doses of carbon monoxide. *J. Publ. Hlth.* **57**, 2012–2022.

Beebee, T. J. (1973). Observations concerning the decline of the British amphibia. *Biol. Conserv.* **5**, 20–24.

Beiningen, K. T. & Ebel, W. F. (1970). Effects of the John Day Dam on dissolved nitrogen concentrations and salmon in the Columbia River, 1968. *Trans. Am. Fish. Soc.* **99**, 664–671.

Bell, R. H. V. (1971). A grazing system in the Serengeti. *Scient. Am.* **225**, 86–93.

Bentley, E. W. (1960). Control of rats in sewers. *Tech. Bull. Minist. Agric. Fish. Fd.*, **10**, 1–22.

Bere, R. (1975). *Mammals of East and Central Africa*. Longman, London.

Bernarde, M. A. (1973). *Our Precarious Habitat.* W. W. Norton, New York.

Bisseru, B. (1967). *Diseases of Man acquired from his Pets*. Heinemann, London.

Blokpoel, H. (1974). Migration of Lesser Snow and Blue Geese in spring across Southern Manitoba. Part 1: Distribution, chronology, directions, numbers, heights and speeds. *Can. Wildl. Serv. Rep. Ser.* **28**, 1–30.

Blood, B. D., Moya, V. & Lelijveld, J. L. (1968). Evaluation of selected drugs for the treatment of canine echinococcosis. *Bull. Wld. Hlth Org.* **39**, 67–72.

Blower, J. (1968). The wildlife of Ethiopia. *Oryx*, **9**, 276–283.

Bond, R. G. & Straub, C. P. (Ed.) (1973–1974). *Handbook of Environmental Control.* 4 vols. CRC Press, Cleveland, Ohio.

Booth, A. H. (1960). *Small Mammals of West Africa.* Longman, London.

Borg, O. A. & Woodruff, A. W. (1973). Prevalence of infective ova of *Toxocara* species in public places. *Brit. med. J.* 470–472.

Bradshaw, A. D. (1970). Plants and industrial waste. *Trans. bot. Soc. Edinb.* **41**, 71–84.

Brooks, E. (1973). Twilight of Brazilian tribes. *Geogrl. Mag., Lond.* **45**, 304–310.

Brough, T. (1968). Recent developments on bird scaring on airfields. In *The Problems of Birds as Pests.* (Ed. by R. K. Murton and E. N. Wright), pp. 29–38. Academic Press, London.

Brown, V. M. & Dalton, R. A. (1970). The acute toxicity to rainbow trout of mixtures of copper, phenol, zinc and nickel. *J. Fish. Biol.* **2**, 211–216.

British Field Sports Society and Council for Nature (undated). *Predatory Birds in Britain.* London.

Brummage, M. K., Edington, J. M. & Edwards, R. W. (1977). *Ecological Problems Associated with Changing Patterns of Hand and Natural Resource Use in Wales.* Welsh Office, Cardiff.

Buckner, C. H. (1967). Avian and mammalian predators of forest insects. *Entomophaga*, **12**, 491–501.

Building Design Partnership, (1970). *Rhondda Valleys Development Plan, a Future for a Former Mining Community.*

Burns, W. (1973). *Noise and Man.* John Murray, London.

Burns, K. N. & Allcroft, R. (1964). *Fluorosis in Cattle. 1. Occurrence and Effects in Industrial Areas of England and Wales* 1954–1957. H.M.S.O., London.

Butterworth, J., Lester, P. & Nickless, G. (1972). Distribution of heavy metals in the Severn Estuary. *Mar. Pollut. Bull.* **3**, 72–74.

Cannon, H. L. (1971). The use of plant indicators in ground water surveys, geologic mapping, and mineral prospecting. *Taxon*, **20**, 227–256.

Catchpole, C. K. & Tydeman, C. F. (1975). Gravel pits as new wetland habitats for the conservation of breeding bird communities. *Biol. Conserv.* **8**, 47–59.

Chestney, R. (1971). Conservation of terns in Norfolk, England. *Biol. Conserv.* **4**, 67–69.

Chisolm, J. J. (1971). Lead poisoning. *Scient. Am.* **224**, 15–23.

Christie, W. J. (1974). Changes in the fish species composition of the Great Lakes. *J. Fish. Res. Bd Can.* **31**, 827–854.

Civic Trust (1965). *The Rhondda Valleys, Proposals for the Transformation of an Environment.* Civic Trust, London.

Clark, J. R. (1969). Thermal pollution and aquatic life. *Scient. Am.* **220**, 19–27.

Clark, T. W. (1976). The black-footed ferret. *Oryx*, **13**, 275–280.

Clay, C. H. (1961). *Design of Fishways and other Fish Facilities.* Department of Fisheries of Canada, Ottawa.

Committee for Environmental Conservation (1972). *Some Effects of Supersonic Flights*, London.

Cook, B. R. & Crewe, W. (1963). The epidemiology of *Echinococcus* infection in Great Britain 1. Abnormal behaviour of sheep in the mining valleys of South Wales and its relation to hydatid disease in man. *Ann. trop. Med. Parasit.* **57**, 150–156.

Cott, H. B. (1969). Tourists and crocodiles in Uganda. *Oryx*, **10**, 153–160.

Council for Nature (1973). *Predatory Mammals in Britain.* London.

Croxall, J. P. (1975). The effect of oil on nature conservation, especially birds. In *Petroleum and the Continental Shelf of North-West Europe. 2. Environmental Protection.* (Ed. by H. A. Cole), pp. 93–101. Applied Science Publishers, Barking, England.

Curtis, L. F., Doornkamp, J. C. & Gregory, K. J. (1965). The description of relief in field studies of soils. *J. Soil Sci.* **16**, 16–30.

Dagg, A. I. (1974). *Canadian Wildlife and Man.* McClelland and Stewart, Toronto.

Damann, K. E. (1973). Drinking water or sewage, is there a difference? *Inaugural Lectures Series*, **9**, 1–22. University of Ife Press, Ile-Ife.

Dasmann, R. F. (1964). *African Game Ranching.* Pergamon Press, London.

Davidson, F. A., Vaughan, E., Hutchinson, S. J. & Pritchard, A. L. (1943). Factors influencing the upstream migration of pink salmon (*Oncorhynchus gorbuscha*). *Ecology*, **24**, 149–168.

Davis, B. N. K. (1976). Wildlife, urbanisation and industry. *Biol. Conserv.* **10**, 249–291.

Davis, R. A. (1970). Control of rats and mice. *Techn. Bull. Minist. Agric. Fish. Fd.* **181**, 1–28.

Department of the Environment (1974a). *New Housing and Road Traffic Noise, a Design Guide for Architects.* H.M.S.O., London.

Department of Environment (1974b). *Report of a River Pollution Survey of England and Wales, 1973.* Vol. 3, H.M.S.O., London.

Department of the Environment (1976a). *Central Unit on Environmental Pollution. Effects of Airborne Sulphur Compounds on Forests and Freshwaters*. H.M.S.O., London.

Department of the Environment (1976b). *Building Research Establishment Report. Predicting Road Traffic Noise*. H.M.S.O., London.

Dorolle, P. (1968). Old plagues in the jet age. International aspects of present and future control of communicable disease. *Brit. med. J.* **4**, 789–792.

Dorset Naturalists' Trust, (1970). *Farming and Wildlife in Dorset*. Report of a Study Conference. Poole.

Douthwaite, R. J. (1974). An endangered population of wattled cranes (*Grus carunculatus*). *Biol. Conserv.* **6**, 134–142.

Drake, M. F. (1973). Foulness: wildlife and the airport. *Ecologist*, **3**, 140–143.

Drummond, D. C. (1966). Rats resistant to warfarin. *New Scient.* **30**, 771–772.

Duffey, E. (1968). An ecological analysis of the spider fauna of sand dunes. *J. Anim. Ecol.* **37**, 641–674.

Duguid, I. M. (1961). Features of ocular infestation by *Toxocara*. *Br. J. Ophthal.* **45**, 789–796.

Edington, J. M., Morgan, P. J. & Morgan, R. A. (1973). Feeding patterns of wading birds on the Gann Flat and river estuary at Dale. *Field Stud.* **3**, 783–800.

Edwards, R. (1971). The polychlorobiphenyls, their occurrence and significance: a review. *Chem. Ind.*, 1340–1348.

Ekman, S. (1953). *Zoogeography of the Sea*. Sidgwick and Jackson, London.

Ellis, J. B. (1974). The Brecon Beacons National Park, conflicts and issues. *Tn Ctry Plann.* **42**, 458–465.

Elton, C. S. (1958). *The Ecology of Invasions by Plants and Animals*. Methuen, London.

Emmerson, B. T. (1970). "Ouch-Ouch" disease, the osteomalacia of cadmium nephropathy. *Ann. intern. Med.* **73**, 854–855,

Ene, J. C. (1963). *Insects and Man in West Africa*. Ibadan University Press, Ibadan.

Farid, M. A. (1975). The Aswan High Dam development project. In *Man-Made Lakes and Human Health*, (Eds. N. F. Stanley and M. P. Alpers), pp. 89–102. Academic Press, London.

Fimriette, N., Fyfe, R. W. & Keith, J. A. (1970). Mercury contamination of Canadian prairie seed eaters and their avian predators. *Can. Fld Nat.* **84**, 269–276.

Fosbrooke, H. (1972). *Ngorongoro: the Eighth Wonder*. Andre Deutsch, London.

Franz, J. M. (1961). Biological control of pest insects in Europe. *A. Rev. Ent.* **6**, 183–200.

Frost, P. G. H., Seigfried, W. R. & Cooper, J. (1976). Conservation of the jackass penguin (*Spheniscus demersus* (L)). *Biol. Conserv.* **9**, 79–99.

Fryer, G. & Iles, T. D. (1972). *The Cichlid Fishes of the Great Lakes of Africa: their Biology and Evolution*. Oliver and Boyd, Edinburgh.

Fyfe, R. W., Campbell, J., Hayson, B. & Hodson, K. (1969). Regional population declines and organochlorine insecticides in Canadian prairie falcon. *Can. Fld Nat.* **83**, 191–200.

Gadgil, R. L. (1969). Tolerance of heavy metals and the reclamation of industrial waste. *J. appl. Ecol.* **6**, 247–259.

Geldreich, E. E. (1967). Faecal coliform concepts in stream pollution. *Water Sewage Works*, **114**, 98–109.

Geldreich, E. E. & Kenner, B. A. (1969). Concepts of faecal streptococci in stream pollution. *J. Wat. Pollut. Control Fed.* **41**, 336–352.

Gilbert, O. L. (1975). Effects of air pollution on landscape and land-use around Norwegian aluminium smelters. *Environ. Pollut.* **8**, 113–121.

Godwin, H. & Walters, S. M. (1967). The scientific importance of Upper Teesdale. *Proc. bot. Soc. Br. Isl.* **6**, 348–351.

Goldsmith, E., Allen, R., Allaby, M., Davoll, J. & Lawrence, S. (1972). *A Blueprint for Survival* (reprinted from the Ecologist). Penguin Books, Harmondsworth.

Goodland, R. J. A. & Irvin, H. S. (1975). *Amazon Jungle: Green Hell to Red Desert? An Ecological Discussion of the Environmental Impact of the Highway Construction Program in the Amazon Basin.* Elsevier, Amsterdam.

Goodman, G. T. & Chadwick, M. J. Ed. (1975). *The Ecology of Resource Degradation and Renewal. Brit. Ecol. Soc. Symp.* **15**. Blackwell, Oxford.

Goodman, G. T. & Chadwick, M. J. Ed. (1975). The Ecology of Resource Degredation and Renewal. *Brit. Ecol. Soc. Symp.* **15**. Blackwell, Oxford.

Gotaas, H. B. (1956). *Composting – Sanitary Disposal and Reclamation of Organic Wastes.* World Health Organisation, Geneva.

Greaves, J. H., Hammond, L. E. & Bathard, A. H. (1968). The control of re-invasion by rats of part of a sewer network. *Ann. appl. Biol.* **62**, 341–351.

Greenwood, H. P. (1967). Blind cave fishes. *Stud. Speleol.* **1**, 262–274.

Grzimek, B. & Grzimek, K. (1964). *Serengeti Shall Not Die.* Fontana-Collins, Glasgow.

Gwynne, M. D. & Bell, R. H. V. (1968). Selection of vegetation components by grazing ungulates in the Serengeti National Park. *Nature*, **220**, 290–393.

Haagen-Smit. (1968). Air conservation. *Scientia, Milan.* **103**, 261–280.

Hammond, P. B. & Aronson, A. L. (1964). Lead poisoning in cattle and horses in the vicinity of a smelter. *Ann. N.Y. Acad. Sci.* **111**, 595–611.

Happold, D. C. D. (1973). *Large Mammals of West Africa.* Longman, London.

Harden-Jones, F. R. (1968). *Fish Migration.* Arnold, London.

Hardisty, M. W., Huggins, R. J., Karter, S. & Sainsbury, M. (1974). Ecological implications of heavy metals in fish from the Severn Estuary. *Mar. Pollut. Bull.* **5**, 12–15.

Harkins, W. D. & Swain, R. W. (1908). The chronic arsenical poisoning of herbivorous animals. *J. Am. Chem. Soc.* **30**, 928–46.

Hawksworth, D. L. & Rose, F. (1970). Qualitative scale for estimating sulphur dioxide air pollution in England and Wales using epiphytic lichens. *Nature*, **227**, 145–148.

Hay, A. (1976a). Toxic cloud over Seveso. *Nature*, **262**, 636–638.

Hay, A. (1976b). Seveso: the aftermath. *Nature*, **263**, 538–540.

Hayes, F. R. (1953). Artificial freshets and other factors controlling the ascent and population of Atlantic salmon in the La Have River, Nova Scotia. *Bull. Fish. Res. Bd Can.* **99**, 1–47.

Heathcote, A., Griffin, D. & Salmon, H. M. (1967). *The Birds of Glamorgan.* Cardiff Naturalists' Society.

Herbert, D. W. M. & Vandyke, J. M. (1964). The toxicity to fish of mixtures of poisons. II Copper-ammonia and zinc-phenol mixtures. *Ann. appl. Biol.* **53**, 415–421.

Highton, R. B. & Van Someren. (1970). The transportation of mosquitos between international airports. *Bull. Wld Hlth Org.* **42**, 334–335.

Hinz, E. (1967). Geschlechts-und Altersunterschiede im Darmhelminthenbefall bei der Bevölkering Südnigerias. *Z. Tropenmed. Parasit.* **18**, 334–342.

Hira, P. R. (1969). Transmission of schistosomiasis in Lake Kariba, Zambia. *Nature*, **224**, 670–672.

Hirst, L. F. (1953). *The Conquest of Plague – a Study of the Evolution of Epidemiology*. Clarendon Press, Oxford.

Hooker, A. V. (1970). Severnside of the future. *Proc. Instn civ. Engrs.* **47**, 337–348.

Hooper, J. H. D. (1964). Bats and the amateur naturalist. *Stud. Speleol.* **1**, 9–15.

Howard, J. A. (1970). Airport on the estuary. *New Scient.* **45**, 302–304.

Hutchinson, T. C. & Whitby, L. M. (1974). Heavy-metal pollution in the Sudbury mining and smelting region of Canada, 1. Soil and vegetation contamination by nickel, copper, and other metals. *Environ. Conserv.* **1**, 123–132.

Hutnik, R. J. & Davis, G. M. (Ed.) (1973). *Ecology and Reclamation of Devastated Land*. 2 vols. Gordon and Breach, New York.

Huxley, J. S. (1961). *The Conservation of Wild Life and Natural Habitats in Central and East Africa*. U.N.E.S.C.O., Paris.

Hyde, H. A. (1961). *Welsh Timber Trees*. National Museum of Wales, Cardiff.

Hynes, H. B. N. (1960). *The Biology of Polluted Waters*. Liverpool University Press, Liverpool.

Ingram, G. C. S. & Salmon, H. M. (1957). The birds of Brecknock. *Brycheiniog*, **3**, 182–259.

Ivens, G. W. (1974). *Eupatorium odoratum* L. in Nigeria. *Pest Articles News. Summ.*, **20**, 76–82.

Jenkins, D., Watson, A. & Miller, G. R. (1964). Predation and red grouse populations. *J. appl. Ecol.* **1**, 183–195.

Jenkins, D., Watson, A. & Miller, G. R. (1967). Population fluctuations in the red grouse, *Lagopus lagopus scoticus*. *J. Anim. Ecol.* **36**, 97–122.

Jenson, S. & Jernelöv, A. (1969). Biological methylation of mercury in aquatic organisms. *Nature*, **223**, 753–754.

Johnson, R. (1975). Impact assessment is on the way. *New Scient.* **68**, 323–325.

Johnston, H. (1971). Reduction of statospheric ozone by nitrogen oxide catalysts from supersonic transport exhaust. *Science N.Y.* **173**, 517–522.

Jonathan, O. M. (1960). Hydatid disease in North Wales. *Br. med. J.* 1246–1253.

Jones, J. W. (1959). *The Salmon*. Collins, London.

Klein, L. (Ed.) (1962). *River Pollution*. 3 vols. Butterworths, London.

Kury, C. B. & Gochfield, M. (1975). Human interference and gull predation in cormorant colonies. *Biol. Conserv.* **8**, 23–34.

Larkin, P. A. & Northcote, T. G. (1969). Fish as indices of eutrophication. In *Eutrophication: Causes, Consequences and Correctives*. pp. 256–273. Nat. Acad. Sci. Washington D.C.

Laws, R. M. (1969). The Tsavo Research Project. *J. Reprod. Fertil. Suppl.* **6**, 495–531.

Laws, R. M. (1970). Elephants as agents of habitat and landscape change in East Africa. *Oikos*, **21**, 1–15.

Lawson, R. M. (1963). The economic organisation of the *Egeria* fishing industry on the River Volta. *Proc. malac. Soc. Lond.* **35**, 273–287.

Learner, M. A., Williams, R., Harcup, M. & Hughes, B. D. (1971). A survey of the macro-fauna of the River Cynon, a polluted tributary of the River Taff (South Wales). *Freshwat. Biol.* **1**, 339–367.

Lemberg, K. (1975). Finding sites for major airports: the experience of Copenhagen. In *Airports and the Environment.* pp. 99–115. O.E.C.D., Paris.

Little, E. C. S. (1965). The discovery of *Salvinia auriculata* on the Congo. *Nature*, **208**, 1111–1112.

Little, P. & Martin, M. H. (1974). Biological monitoring of heavy metal pollution. *Environ. Pollut.* **6**, 1–19.

Lloyd, R. J. (1970). *Countryside Recreation, the Ecological Implications.* Lindsey County Council.

Lowe, V. P. W. (1969). Population dynamics of the red deer (*Cervus elaphus* L.) on Rhum. *J. Anim. Ecol.*, **38**, 425–57.

Lowe-McConnell, R. H. (Ed.) (1966). *Man-made Lakes.* Academic Press, London.

Lucas, D. (1974). Pollution control by tall chimneys. *New Scient.* **63**, 790–791.

Luther, H. & Rzóska, J. (1971). *Project Aqua.* IBP Handbook No. 21. Blackwell, Oxford.

MacArthur, R. H. & MacArthur, J. W. (1961). On bird species diversity. *Ecology*, **42**, 594–598.

MacFarland, C. G. & Reeder, W. G. (1975). Breeding, raising and restocking of giant tortoises (*Geochelone elephantopus*) in the Galapagos Islands. In *Breeding Endangered Species in Captivity.* (Ed. by R. D. Martin), pp. 13–37. Academic Press, London.

MacFarland, C. G., Villa, J. & Toro, B. (1974a). The Galapagos giant tortoises (*Geochelone elephantopus*). 1. Status of the surviving populations. *Biol. Conserv.* **6**, 118–133.

MacFarland, C. G., Villa, J. & Toro, B. (1974b). The Galapagos giant tortoises (*Geochelone elephantopus*) II. Conservation methods. *Biol. Conserv.* **6**, 198–212.

Maddox, J. (1972). *The Doomsday Syndrome.* Macmillan, London.

Manasseh, L. & partners (1975). *Snowdon Summit.* Countryside Commission, Cheltenham.

Marshall, B. E. & Falconer, A. C. (1973). Physico-chemical aspects of Lake McIlwaine (Rhodesia) a eutrophic tropical impoundment. *Hydrobiologia*, **42**, 45–62.

Mattingly, P. F. (1969). *The Biology of Mosquito-borne Disease.* Allen and Unwin, London.

McClachlan, A. J. (1970). Submerged trees as a substrate for benthic fauna in the recently created Lake Kariba (Central Africa). *J. appl. Ecol.* **7**, 253–266.

McClachlan, A. J. (1974). Development of some lake ecosystems in tropical Africa, with special reference to the invertebrates. *Biol Rev.* **49**, 365–397.

McHarg, I. L. (1969). *Design with Nature.* Natural History Press, New York.

Meadows, D. L., Meadows, D. H., Randers, J. & Behrens, W. W. (1972). *The Limits to Growth.* Earth Island, London.

Mech, D. (1972). *The Wolf: The Ecology and Behaviour of an Endangered Species.* Constable, London.

Medical Research Council (1959). Committee on Bathing Beach Contamination, Sewage Contamination of Coastal Bathing Waters in England and Wales. Med. Res. Council. Mem. 37, H.M.S.O., London.

Menser, H. A. & Heggestad, H. E. (1966). Ozone and sulfur dioxide synergism: injury to tobacco plants. *Science, N.Y.* **153**, 424–425.

Ministry of Housing and Local Government (1969). *The Bacteriological Examination of Water Supplies*. 4th ed. H.M.S.O., London.

Mitchell, D. S. (Ed.), (1974). *Aquatic Vegetation and its Use and Control.* U.N.E.S.C.O., Paris.

Mitchell, D. S. & Thomas, P. A. (1972). *Ecology of Water Weeds in the Neotropics*, U.N.E.S.C.O., Paris.

Monath, T. P. (1975). Lassa fever: review of epidemiology and epizootiology. *Bull. Wld Hlth Org.* **52**, 577–592.

Mountjoy, A. B. (1975). *Industrialisation and Underdeveloped Countries.* (4th edn.) Hutchinson, London.

Munro, J. L. (1966). A limnological survey of Lake McIlwaine, Rhodesia. *Hydrobiologia*, **28**, 281–308.

Murray, M. M. & Wilson, D. C. (1946). Fluorine hazards with special reference to some social consequences of industry processes. *Lancet*, **2**, 821–824.

Murton, R. K. (1971). *Man and Birds*. Collins, London.

Myers, N. (1973). Tsavo National Park, Kenya and its elephants: an interim appraisal. *Biol. Conserv.*, **5**, 123–132.

Myres, M. T. & Cannings, S. R. (1971). A Canada Goose migration through the southern interior of British Columbia. *Can. Wildl. Serv. Rep. Ser.* **14**, 23–34.

Nair, N. B. & Saraswathy, M. (1971). The biology of wood-boring teredinid molluscs. *Adv. mar. Biol.* **9**, 335–509.

Napier, J. R. & Napier, P. H. (1967). *A Handbook of Living Primates*. Academic Press, London.

Nash, C. E. (1968). Power stations as sea farms. *New Scient.* **40**, 367–369.

Naylor, E. (1965a), Biological effects of a heated effluent in docks at Swansea, S. Wales. *Proc. zool. Soc. Lond.* **144**, 253–268.

Naylor, E. (1965b). Effects of heated effluents upon marine and estuarine organisms. *Adv. mar. Biol.* **3**, 63–103.

Nickless, G., Stenner, R. & Terrille, N. (1972). Distribution of cadmium, lead and zinc in the Bristol Channel. *Mar. Pollut. Bull.* **3**, 188–190.

Nilsson, L. (1972). Local distribution, food choice and food consumption of diving ducks on a South Swedish Lake. *Oikos* **23**, 82–91.

Noise Advisory Council (1973). *A Guide to Noise Units*. H.M.S.O., London.

Noise Advisory Council (1974). *Aircraft Engine Noise Research.* H.M.S.O., London.

Obeng, L. E. (Ed.) (1969). *Man-made Lakes, the Accra Symposium.* Ghana Universities Press.

Olney, P. J. S. (1963a). The food and feeding habits of the teal, *Anas crecca crecca* L. *Proc. zool. Soc. Lond.* **140**, 169–210.

Olney, P. J. S. (1963b). The food and feeding habits of tufted duck, *Aythya fuligula*. *Ibis*, **105**, 55–62.

Olney, P. J. S. (1968). The food and feeding habits of the pochard, *Aythya ferina*. *Biol. Conserv.* **1**, 71–76.

Olney, P. J. S. & Mills, D. H. (1963). The food and feeding habits of goldeneye, *Bucephala clangula*, in Great Britain. *Ibis*, **105**, 293–300.

Ogilvie, M. A. & Matthews, G. V. T. (1969). Brent geese, mudflats and man. *Wildfowl*, **20**, 119–125.

Owen, M. (1973). The winter feeding of wigeon at Bridgwater Bay, Somerset. *Ibis*, **115**, 227–243.

Pannell, J. P. M., Johnson, A. E. & Raymont, J. E. G. (1962). An investigation into the effects of warmed water from Marchwood Power Station into Southampton Water. *Proc. Instn. Civ. Engrs.* **23**, 35–62.

Paperna, I. (1969). Evolution of the shoreline, aquatic weeds, snails and bilharzia transmission in the newly formed Volta Lake. *Verh. int. Ver. Limnol.* **17**, 282–283.

Paperna, I. (1970). Study of an outbreak of schistosomiasis in the newly formed Volta Lake, Ghana. *Z. Tropenmed. Parasit.* **21**, 411–425.

Patterson, D. J. & Henein, N. A. (1972). *Emissions from Combustion Engines and their Control.* Ann Arbor Science Publishers, Ann Arbor, Michigan.

Pearsall, W. H. (1957). Report on an ecological survey of the Serengeti National Park, Tanganyika. *Oryx*, **4**, 71–136.

Penny, M. (1968). Endemic birds of the Seychelles. *Oryx*, **9**, 267–275.

Petr, T. (1968). Distribution, abundance and food of commercial fish in the Black Volta and the Volta man-made lake in Ghana during its first period of filling (1964–1966). I Mormyridae. *Hydrobiologia*, **32**, 417–448.

Petr, T. (1970). Macro-invertebrates of flooded trees in the man-made Volta Lake (Ghana) with special reference to the burrowing mayfly *Povilla adusta* Navas. *Hydrobiologia*, **36**, 399–418.

Petr, T. (1971). Lake Volta – a progress report. *New Scient.* **49**, 178–182.

Pigott, C. D. (1956). The vegetation of Upper Teesdale in the North Pennines. *J. Ecol.* **44**, 545–586.

Pimlott, D. H. (1967). Wolf predation and ungulate populations. *Am. Zool.* **7**, 267–278

Pollock, N. C. (1969). Some observations on game ranching in Southern Africa. *Biol. Conserv.*, **2**, 18–24.

Prater, A. J. (1974). *Birds of Estuaries Enquiry 1972–73.* British Trust for Ornithology. Tring, Herts.

Prescott, G. W. (1969). *The Algae: a Review.* Nelson, London.

Preston, A., Jefferies, D. F., Dutton, J. N. R., Harvey, B. R. & Steel, A. K. (1972). British Isles coastal waters: the concentration of selected heavy metals in sea water, suspended matter and biological indicators – a pilot study. *Environ. Pollut.* **3**, 69–82.

Pryde, P. R. (1972). *Conservation in the Soviet Union.* Cambridge University Press, London.

Pyatt, F. B. (1970). Lichens as indicators of air pollution in a steel-producing town in South Wales. *Environ. Pollut.* **1**, 45–56.

Pyefinch, K. A. (1947). The biology of ship fouling. *New Biol.* **3**, 128–148.

Pyefinch, K. A. (1966). Hydro-electric schemes in Scotland. Biological problems and effects on salmonid fisheries. In *Man-made Lakes.* (Ed. by R. H. Lowe-McConnell), pp. 139–147. Academic Press, London.

Racey, P. A. & Stebbings, R. E. (1972). Bats in Britain – a status report. *Oryx*, **11**, 319–327.

Randhawa, H. S., Clayton, Y. M. & Riddell, R. W. (1965). Isolation of *Cryptococcus neoformans* from pigeon habitats in London. *Nature*, **208**, 801.

Rao, T. R., Trpis, M., Gillett, J. D., Teesdale, C. & Tonn, R. J. (1973). Breeding places and seasonal incidence of *Aedes aegypti*, as assessed by the single larva survey method. *Bull. Wld Hlth Org.* **48**, 615–622.

Rasmussen, D. I. (1941). Biotic communities of the Kaibab Plateau. *Ecol. Monogr.* **3**, 229–275.

Reinhard, H. (1975). Abating noise: a Swiss approach. In *Airports and the Environment.* pp. 205–215. O.E.C.D., Paris.

Richards, P. W. (1939). Ecological studies on the rain forest of Southern Nigeria. I. The structure and floristic composition of the primary forest. *J. Ecol.* **27**, 1–61.

Richards, P. W. (1952). *The Tropical Rain Forest.* Cambridge University Press, London.

Ricks, G. R. & Williams, R. J. H. (1974). Effects of atmospheric pollution on deciduous woodland. Part 2. Effects of particulate matter upon stomatal diffusion resistance in leaves of *Quercus petraea* (Mattuschka) Leibl. *Environ. Pollut.* **6**, 87–109.

Ricks, G. R. & Williams, R. J. H. (1975). Effects of atmospheric pollution on deciduous woodland. Part 3: Effects on photosynthetic pigments of leaves of *Quercus petraea* (Mattuschka) Leibl. *Environ. Pollut.* **8**, 97–106.

Robinson, P. T. (1971). Wildlife trends in Liberia and Sierra Leone. *Oryx*, **11**, 117–122.

Royal College of Physicians (1970). *Air Pollution and Health.* Pitman, London.

Rudd, R. L. (1964). *Pesticides and the Living Landscape.* University of Wisconsin Press. Madison.

Saul, E. K. (1967). Birds and aircraft: a problem at Auckland's new international airport. *Jl. R. aeronaut. Soc.* **71**, 366–376.

Schindler, D. W. & Fee, E. J. (1974). Experimental lakes area: whole-lake experiments in eutrophication. *J. Fish. Res. Bd Can.* **31**, 937–953.

Schmitt, N., Brown, G., Devlin, E. F., Larsen, A. A., McCausland, E. D. & Saville, J. M. (1971). Lead poisoning in horses. *Archs envir. Hlth.* **23**, 185–197.

Schofield, E.K. (1973). Galapagos flora, the threat of introduced plants. *Biol Conserv.* **5**, 48–51.

Schofield, J. M. (1967). Human impact on the fauna, flora and natural features of Gibraltar Point. In *The Biotic Effects of Public Pressure on the Environment.* (Ed. by E. Duffey), pp. 106–111. Nature Conservancy.

Schulte, J. H. (1963). Effects of mild carbon monoxide intoxication. *Archs. envir. Hlth.* **7**, 524–530.

Shaw, T. L. (1974). Tidal energy from the Severn Estuary. *Nature*, **249**, 730–733.

Sherman, R. (1966). The insect jet set. *New Scient.* **30**, 729–732.

Shrewsbury, J. F. D. (1971). *A History of Bubonic Plague in the British Isles.* Cambridge University Press, London.

Singh, D. (1967). The *Culex pipiens fatigans* problem in South-East Asia with special reference to urbanisation. *Bull. Wld Hlth Org.* **37**, 239–243.

Smith, S. H. (1968). Species succession and fishery exploitation in the Great Lakes. *J. Fish. Res. Bd Can.* **25**, 667–693.

Solomon, D. J. (1973). Evidence for pheromone-influenced homing by migrating Atlantic salmon, *Salmo salar* L. *Nature*, **244**, 231–232.

Soper, F. L. & Wilson, D. B. (1943). *Anopheles gambiae in Brazil 1930–1940.* Rockefeller Foundation, New York.

Stanley, N. F. & Alpers, M. P. (Ed.) (1975). *Man-made Lakes and Human Health.* Academic Press, London.

Stebbings, R. E. (1966). Bats under stress. *Stud. Speleol.* **1**, 168–173.

Stevenson, A. H. (1953). Studies of bathing water quality and health. *Am. J. publ. Hlth.* **43**, 529–538.

Stuart-Harris, C. H. (1965). *Influenza and other Virus Infections of the Respiratory Tract.* Edward Arnold, London.

Sullivan, W. N., Pal, R., Wright, J. W., Azurin, J. C., Okamoto, R., McGuire, J. V. & Waters, R. M. (1972). Worldwide studies on aircraft disinsection at "blocks away". *Bull. Wld Hlth Org.* **46**, 485–491.

Surtees, G. (1971). Urbanisation and the epidemiology of mosquito-borne disease. *Abs. Hyg.* **46**, 121–134.

Sussman, V. H., Lieben, J. & Cleland, J. G. (1959). An air pollution study of a community surrounding a beryllium plant. *Amer. Industr. Hyg. Ass. J.* **20**, 504–508.

Swabey, C. (1970). The endemic flora of the Seychelle Islands and its conservation. *Biol Conserv.* **2**, 171–177.

Tadano, T. & Brown, A. W. A. (1966). Development of resistance to various insecticides in *Culex pipiens fatigans* Wiedemann. *Bull. Wld Hlth Org.* **35**, 189–201.

Talbot, L. M., Payne, W. J. A., Ledger, H. P., Verdcourt, L. D. & Talbot, M. H. (1965). The meat production potential of wild animals in Africa. A review of biological knowledge. *Tech. Commun. Commonw. Bur. Anim. Breed. Genet.* **16**, 1–42.

Thearle, R. J. P. (1968). Urban bird problems. In *The Problems of Birds as Pests.* (Ed. by R. K. Murton and E. N. Wright), pp. 181–97. Academic Press, London.

Thomas, G. J. (1972). A review of gull damage and management methods at nature reserves. *Biol. Conserv.* **4**, 117–127.

Tingey, D. T., Reinert, R. A., Dunning, J. A. & Heck, W. W. (1971). Vegetation injury from the interaction of nitrogen dioxide and sulphur dioxide. *Phytopathology*, **61**, 1506–1511.

Tucker, A. (1972). *The Toxic Metals.* Earth Island, London.

Upton, M. (1967). Agriculture in South-Western Nigeria. *Development Studies*, **3**, University of Reading.

Vale, T. R. (1975). Ecology and environmental issues of the Sierra Redwood (*Sequoiadendron giganteum*), now restricted to California. *Environ. Conserv.* **2**, 179–188.

Valery, N. (1975). SSTs are clean – in small numbers. *New Scient.* **68**, 19–21.

Vallentyne, J. R. (1974). *The Algal Bowl – Lakes and Man.* Department of the Environment, Fisheries and Marine Service. Ottawa.

Varley, M. E. (1967). *British Freshwater Fishes, Factors Affecting their Distribution.* Fishing News (Books) Ltd., London.

Waddy, B. B. (1975). Research into the health problems of man-made lakes, with special reference to Africa. *Trans. R. Soc. trop. Med. Hyg.* **69**, 39–50.

Waldbott, G. L. (1973). *Health Effects of Environmental Pollutants.* C. V. Mosby, Saint Louis.

Waldbott, G. L. & Cecilioni, V. A. (1969). "Neighbourhood" fluorosis. *Fluoride*, **2**, 206–213.

Wallace, A. R. (1876). *The Geographical Distribution of Animals*. 2 vols. MacMillan, London.

Walther, F. R. (1969). Flight behaviour and avoidance of predators in Thomson's gazelle (*Gazella thomsoni* Guenther 1884). *Behaviour*, **34**, 184–221.

Welsh Office (1967). *Wales: The Way Ahead*. Cmnd. 3334. H.M.S.O., London.

West, G. P. (1972). *Rabies in Animals and Man*. David-Charles, Newton Abbot.

White, E. (1973). Zambia's Kafue hydroelectric scheme and its biological problems. In *Man-Made Lakes: their Problems and Environmental Effects* (Ed. by W. C. Ackerman, G. F. White and E. B. Worthington), pp. 620–628. Geophysical Monograph Series 17.

Wilcocks, C. & Manson-Bahr, P. E. C. (Ed.). (1972). *Manson's Tropical Diseases*. 17th ed. Bailliere Tindall, London.

Willard, B. E. & Marr, J. W. (1970). Effects of human activities on alpine tundra ecosystems in Rocky Mountain National Park, Colorado. *Biol. Conserv.* **2**, 257–265.

Willard, B. E. & Marr, J. W. (1971). Recovery of alpine tundra under protection after damage by human activities in the Rocky Mountains of Colorado. *Biol. Conserv.* **3**, 181–190.

Williams, M. A. (1966). Further investigations into bacterial and algal populations of caves in South Wales. *Int. J. Speleol.* **2**, 389–395.

Wodzicki, K. (1973). Prospects for biological control of rodent populations. *Bull. Wld Hlth Org.* **48**, 461–467.

Wolman, A. (1965). The metabolism of cities. *Scient. Am.* **213**, 178–190.

Woodruff, A. W. (1970). Toxocariasis. *Brit. med. J.* 663–669.

Woodruff, A. W., Bisseru, B. & Bowe, J. C. (1966). Infection with animal helminths as a factor in causing poliomyelitis and epilepsy. *Brit. med. J.* 1576–1579.

World Health Organisation. (1964). Soil-transmitted helminths. *Techn. Rep. Ser.* **277**, 1–70.

World Health Organisation (1969). Urban air pollution with particular reference to motor vehicles. *Techn. Rep. Ser.* **410**, 1–53.

World Health Organisation (1971). *International Standards for Drinking-Water* (3rd edn.) Geneva.

World Health Organisation (1972a). Air quality criteria and guides for urban air pollutants. *Techn. Rep. Ser.* **506**, 1—35.

World Health Organisation (1972b). *Vector Control in International Health*. Geneva.

World Health Organisation (1974a). *International Health Regulations (1969)* 2nd annotated edition, Geneva.

World Health Organisation (1974b). Fish and shellfish hygiene. *Techn. Rep. Ser.* **550**, 1–62.

World Meteorological Organisation (1976). Statement on modification of ozone layer due to human activities. *Environ. Conserv.* **3**, 68–70.

Young, J. S. & Gibson, C. I. (1973). Effect of thermal effluent on migrating menhaden. *Mar. Pollut. Bull.* **4**, 94–95.

# Index